Use R!

Series Editors:
Robert Gentleman   Kurt Hornik   Giovanni Parmigiani

For other titles published in this series, go to
http://www.springer.com/series/6991

J.O. Ramsay · Giles Hooker · Spencer Graves

Functional Data Analysis
with R and MATLAB

 Springer

J.O. Ramsay
2748, Howe Street
Ottawa, ON K2B 6W9
Canada
ramsay@psych.mcgill.ca

Giles Hooker
Department of Biological Statistics
& Computational Biology
Cornell University
1186, Comstock Hall
Ithaca, NY 14853
USA
gjh27@cornell.edu

Spencer Graves
Productive Systems Engineering
751, Emerson Ct.
San Jose, CA 95126
USA
spencer.graves@prodsyse.com

Series Editors:

Robert Gentleman
Program in Computational Biology
Division of Public Health Sciences
Fred Hutchinson Cancer
Research Center
1100, Fairview Avenue, N. M2-B876
Seattle, Washington 98109
USA

Kurt Hornik
Department of Statistik
and Mathematik
Wirtschaftsuniversität
Wien Augasse 2-6
A-1090 Wien
Austria

Giovanni Parmigiani
The Sidney Kimmel
Comprehensive Cancer Center
at Johns Hopkins University
550, North Broadway
Baltimore, MD 21205-2011
USA

ISBN 978-0-387-98184-0   e-ISBN 978-0-387-98185-7
DOI 10.1007/978-0-387-98185-7
Springer Dordrecht Heidelberg London New York

Library of Congress Control Number: 2009928040

Printed on acid-free paper

Springer is part of Springer Science+Business Media (www.springer.com)

Preface

This contribution to the useR! series by Springer is designed to show newcomers how to do functional data analysis in the two popular languages, Matlab and R. We hope that this book will substantially reduce the time and effort required to use these techniques to gain valuable insights in a wide variety of applications.

We also hope that the practical examples in this book will make this learning process fun, interesting and memorable. We have tried to choose rich, real-world problems where the optimal analysis has yet to be performed. We have found that applying a spectrum of methods provides more insight than any single approach by itself. Experimenting with graphics and other displays of results is essential.

To support the acquisition of expertise, the "scripts" subdirectory of the companion fda package for R includes files with names like "fdarm-ch01.R", which contain commands in R to reproduce virtually all of the examples (and figures) in the book. This can be found on any computer with R and fda installed using `system.file('scripts', package='fda')`. The Matlab code is provides as part of the fda package for R. From within R, it can be found using `system.file('Matlab', package='fda')`. It also can obtained by downloading the `.tar.gz` version of the fda package for R from the Comprehensive R Archive Network (CRAN, www.r-project.org), unzipping it and looking for the `inst/Matlab` subdirectory.

The contents of a book are fixed by schedules for editing and printing. These script files are not similarly constrained. Thus, in some cases, the script files may perform a particular analysis differently from how it is described in the book. Such differences will reflect improvements in our understanding of preferred ways of performing the analysis described in the book. The web site www.functionaldata.org is a resource for ongoing developments of software, new tools and current events.

The support for two languages is perhaps a bit unusual in this series, but there are good reasons for this. Matlab is expensive for most users, but its for capacity modeling dynamical systems and other engineering applications has been critical in the development of today's fda package, especially in areas such chemical engineering where functional data are the rule rather than the exception and where Matlab is widely used. On the other hand, the extendibility of R, the easy interface with lower-level languages, and above all its cost explain its popularity in many fields served by statisticians, students and new researchers. We hope that we can help many of our readers to appreciate the strengths of each language, so as to invest wisely later on. Secondarily, we hope that any user of either language wanting to learn the other can benefit from seeing the same analyses done in both languages.

As with most books in this useR! series, this is not the place to gain enough technical knowledge to claim expertise in functional data analysis nor to develop new tools. But we do hope that some readers will find enough of value here to want to turn to monographs on functional data analysis already published, such as Ramsay and Silverman (2005), and to even newer works.

We wish to end this preface by thanking our families, friends, students, employers, clients and others who have helped make us what we are today and thereby contributed to this book and to our earlier efforts. In particular, we wish to thank John Kimmel of Springer for organizing this series and inviting us to create this book.

James Ramsay, McGill University
Giles Hooker, Cornell University
Spencer Graves, San Jose, CA

Contents

Chapter 1
Introduction to Functional Data Analysis

The main characteristics of functional data and of functional models are introduced. Data on the growth of girls illustrate samples of functional observations, and data on the US nondurable goods manufacturing index are an example of a single long multilayered functional observation. Data on the gait of children and handwriting are multivariate functional observations. Functional data analysis also involves estimating functional parameters describing data that are not themselves functional, and estimating a probability density function for rainfall data is an example. A theme in functional data analysis is the use of information in derivatives, and examples are drawn from growth and weather data. The chapter also introduces the important problem of registration: aligning functional features.

The use of code is not taken up in this chapter, but R code to reproduce virtually all of the examples (and figures) appears in files "fdarm-ch01.R" in the "scripts" subdirectory of the companion "fda" package for R, but without extensive explanation in this chapter of why we used a specific command sequence.

1.1 What Are Functional Data?

1.1.1 Data on the Growth of Girls

Figure 1.1 provides a prototype for the type of data that we shall consider. It shows the heights of 10 girls measured at a set of 31 ages in the Berkeley Growth Study (Tuddenham and Snyder, 1954). The ages are not equally spaced; there are four measurements while the child is one year old, annual measurements from two to eight years, followed by heights measured biannually. Although great care was taken in the measurement process, there is an average uncertainty in height values of at least three millimeters. Even though each record is a finite set of numbers, their values reflect a smooth variation in height that could be assessed, in principle, as

J.O. Ramsay et al., *Functional Data Analysis with R and MATLAB*, Use R,
DOI: 10.1007/978-0-387-98185-7_1,
© Springer Science + Business Media, LLC 2009

often as desired, and is therefore a height *function*. Thus, the data consist of a sample of 10 *functional* observations $\texttt{Height}_i(t)$.

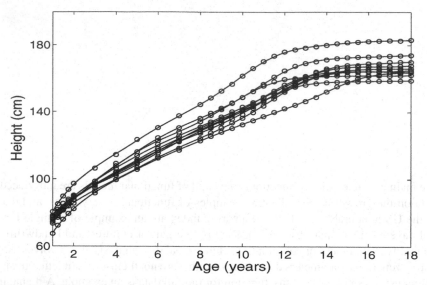

Fig. 1.1 The heights of 10 girls measured at 31 ages. The circles indicate the unequally spaced ages of measurement.

There are features in these data too subtle to see in this type of plot. Figure 1.2 displays the acceleration curves $D^2\texttt{Height}_i$ estimated from these data by Ramsay et al. (1995a) using a technique discussed in Chapter 5. We use the notation D for differentiation, as in

$$D^2\texttt{Height} = \frac{d^2\texttt{Height}}{dt^2}.$$

The pubertal growth spurt shows up as a pulse of strong positive acceleration followed by sharp negative deceleration. But most records also show a bump at around six years that is termed the midspurt. We therefore conclude that some of the variation from curve to curve can be explained at the level of certain derivatives. The fact that derivatives are of interest is further reason to think of the records as functions rather than vectors of observations in discrete time.

The ages are not equally spaced, and this affects many of the analyses that might come to mind if they were. For example, although it might be mildly interesting to correlate heights at ages 9, 10 and 10.5, this would not take account of the fact that we expect the correlation for two ages separated by only half a year to be higher than that for a separation of one year. Indeed, although in this particular example the ages at which the observations are taken are nominally the same for each girl, there is no real need for this to be so. In general, the points at which the functions are observed may well vary from one record to another.

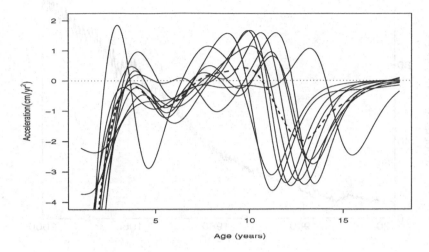

Fig. 1.2 The estimated accelerations of height for 10 girls, measured in centimeters per year per year. The heavy dashed line is the cross-sectional mean and is a rather poor summary of the curves.

The replication of these height curves invites an exploration of the ways in which the curves vary. This is potentially complex. For example, the rapid growth during puberty is visible in all curves, but both the timing and the intensity of pubertal growth differ from girl to girl. Some type of principal components analysis would undoubtedly be helpful, but we must adapt the procedure to take account of the unequal age spacing and the smoothness of the underlying height functions.

It can be important to separate variation in *timing* of significant growth events, such as the pubertal growth spurt, from variation in the *intensity* of growth. We will look at this in detail in Chapter 8 where we consider *curve registration*.

1.1.2 Data on US Manufacturing

Not all functional data involve independent replications; we often have to work with a single long record. Figure 1.3 shows an important economic indicator: the nondurable goods manufacturing index for the United States. Data like these often show variation as multiple levels.

There is a tendency for the index to show geometric or exponential increase over the whole century, and plotting the logarithm of the data in Figure 1.4 makes this trend appear linear while giving us a better picture of other types of variation. At a finer scale, we see departures from this trend due to the depression, World War II, the end of the Vietnam War and other more localized events. Moreover, at an

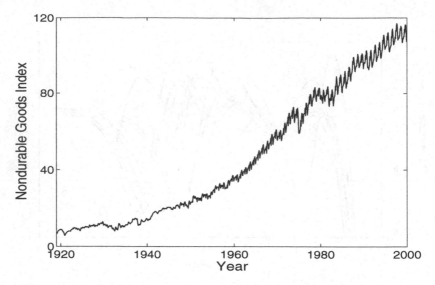

Fig. 1.3 The monthly nondurable goods manufacturing index for the United States.

even finer scale, there is a marked annual variation, and we can wonder whether this *seasonal trend* itself shows some longer-term changes. Although there are no independent replications here, there is still a lot of repetition of information that we can exploit to obtain stable estimates of interesting curve features.

1.1.3 Input/Output Data for an Oil Refinery

Functional data also arise as input/output pairs, such as in the data in Figure 1.5 collected at an oil refinery in Texas. The amount of a petroleum product at a certain level in a distillation column or cracking tower, shown in the top panel, reacts to the change in the flow of a vapor into the tray, shown in the bottom panel, at that level. How can we characterize this dependency? More generally, what tools can we devise that will show how a system responds to changes in critical input functions as well as other covariates?

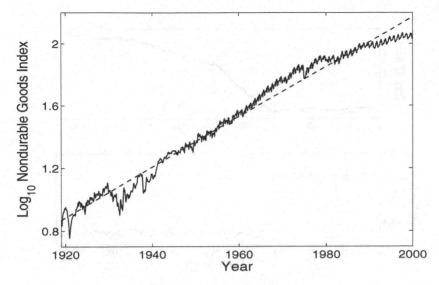

Fig. 1.4 The logarithm of the monthly nondurable goods manufacturing index for the United States. The dashed line indicates the linear trend over the whole time period.

1.2 Multivariate Functional Data

1.2.1 Data on How Children Walk

Functional data are often multivariate. Our third example is in Figure 1.6. The Motion Analysis Laboratory at Children's Hospital, San Diego, CA, collected these data, which consist of the angles formed by the hip and knee of each of 39 children over each child's gait cycle. See Olshen et al. (1989) for full details. Time is measured in terms of the individual gait cycle, which we have translated into values of t in $[0, 1]$. The cycle begins and ends at the point where the heel of the limb under observation strikes the ground. Both sets of functions are periodic and are plotted as dotted curves somewhat beyond the interval for clarity. We see that the knee shows a two-phase process, while the hip motion is single-phase. It is harder to see how the two joints interact: The figure does not indicate which hip curve is paired with which knee curve. This example demonstrates the need for graphical ingenuity in functional data analysis.

Figure 1.7 shows the gait cycle for a single child by plotting knee angle against hip angle as time progresses round the cycle. The periodic nature of the process implies that this forms a closed curve. Also shown for reference purposes is the same relationship for the average across the 39 children. An interesting feature in this plot is the cusp occurring at the heel strike as the knee momentarily reverses its extension to absorb the shock. The angular velocity is clearly visible in terms of the spacing between numbers, and it varies considerably as the cycle proceeds.

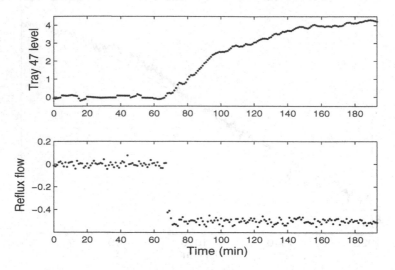

Fig. 1.5 The top panel shows 193 measurements of the amount of petroleum product at tray level 47 in a distillation column in an oil refinery. The bottom panel shows the flow of a vapor into that tray during an experiment.

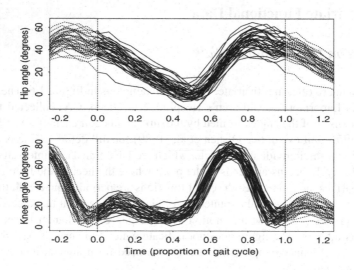

Fig. 1.6 The angles in the sagittal plane formed by the hip and knee as 39 children go through a gait cycle. The interval $[0, 1]$ is a single cycle, and the dotted curves show the periodic extension of the data beyond either end of the cycle.

The child whose gait is represented by the solid curve differs from the average in two principal ways. First, the portion of the gait pattern in the C–D part of the cycle shows an exaggeration of movement relative to the average. Second, in the part of the cycle where the hip is most bent, this bend is markedly less than average; interestingly, this is not accompanied by any strong effect on the knee angle. The overall shape of the cycle for this particular child is rather different from the average. The exploration of variability in these functional data must focus on features such as these.

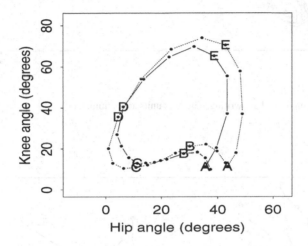

Fig. 1.7 Solid line: The angles in the sagittal plane formed by the hip and knee for a single child plotted against each other. Dotted line: The corresponding plot for the average across children. The points indicate 20 equally spaced time points in the gait cycle. The letters are plotted at intervals of one fifth of the cycle with A marking the heel strike.

1.2.2 Data on Handwriting

Multivariate functional data often arise from tracking the movements of points through space, as illustrated in Figure 1.8, where the X-Y coordinates of 20 samples of handwriting are superimposed. The role of time is lost in plots such as these, but can be recovered to some extent by plotting points at regular time intervals.

Figure 1.9 shows the first sample of the writing of "statistical science" in simplified Chinese with gaps corresponding to the pen being lifted off the paper. Also plotted are points at 120-millisecond intervals; many of these points seem to coincide with points of sharp curvature and the ends of strokes.

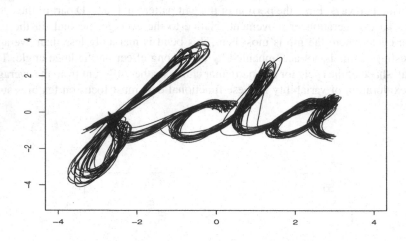

Fig. 1.8 Twenty samples of handwriting. The axis units are centimeters.

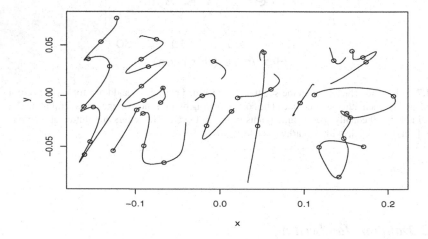

Fig. 1.9 The first sample of writing "statistical science" in simplified Chinese. The plotted points correspond to 120-millisecond time steps.

Finally, in this introduction to types of functional data, we must not forget that they may come to our attention as full-blown functions, so that each record may consist of functions observed, for all practical purposes, everywhere. Sophisticated online sensing and monitoring equipment now routinely used in research in fields such as medicine, seismology, meteorology and physiology can record truly functional data.

1.3 Functional Models for Nonfunctional Data

The data examples above seem to deserve the label "functional" since they so clearly reflect the smooth curves that we assume generated them. Beyond this, functional data analysis tools can be used for many data sets that are not so obviously functional.

Consider the problem of estimating a probability density function p to describe the distribution of a sample of observations $x_1, \ldots, x_n$. The classic approach to this problem is to propose, after considering basic principles and closely studying the data, a *parametric model* with values $p(x|\theta)$ defined by a fixed and usually small number of parameters in the vector θ. For example, we might consider the normal distribution as appropriate for the data, so that $\theta = (\mu, \sigma^2)'$. The parameters themselves are usually chosen to be descriptors of the shape of the density, as in location and spread for the normal density, and are therefore the focus of the analysis.

But suppose that we do not want to assume in advance one of the many textbook density functions. We may feel, for example, that the application cannot justify the assumptions required for using any of the standard distributions. Or we may see features in histograms and other graphical displays that seem not to be captured by any of the most popular distributions. *Nonparametric density* estimation methods assume only smoothness, and permit as much flexibility in the estimated $p(x)$ as the data require or the data analyst desires. To be sure, parameters are often involved, as in the density estimation method of Chapter 5, but the number of parameters is not fixed in advance of the data analysis, and our attention is focused on the density function p itself, not on parameter estimates. Much of the technology for estimation of smooth *functional parameters* was originally developed and honed in the density estimation context, and Silverman (1986) can be consulted for further details.

Psychometrics or mental test theory also relies heavily on functional models for seemingly nonfunctional data. The data are usually zeros and ones indicating unsuccessful and correct answers to test items, but the model consists of a set of *item response functions*, one per test item, displaying the smooth relationship between the probability of success on an item and a presumed latent ability continuum. Figure 1.10 shows three such functional parameters for a test of mathematics estimated by the functional data analytic methods reported in Rossi et al. (2002).

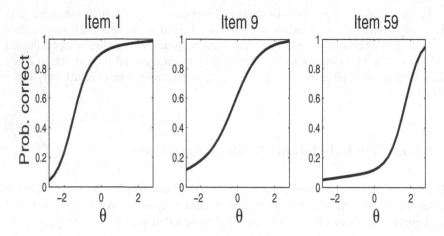

Fig. 1.10 Each panel shows an item response function relating an examinee's position θ on a latent ability continuum to the probability of a correct response to an item in a mathematics test.

1.4 Some Functional Data Analyses

Data in many fields come to us through a process naturally described as functional. Consider Figure 1.11, where the mean temperatures for four Canadian weather stations are plotted as smooth curves. Montreal, with the warmest summer temperature, has a temperature pattern that appears to be nicely sinusoidal. Edmonton, with the next warmest summer temperature, seems to have some distinctive departures from sinusoidal variation that might call for explanation. The marine climate of Prince Rupert is evident in the small amount of annual variation in temperature. Resolute has bitterly cold but strongly sinusoidal temperatures.

One expects temperature to be periodic and primarily sinusoidal in character and over the annual cycle. There is some variation in the timing of the seasons or phase, because the coldest day of the year seems to be later in Montreal and Resolute than in Edmonton and Prince Rupert. Consequently, a model of the form

$$\texttt{Temp}_i(t) \approx c_{i1} + c_{i2}\sin(\pi t/6) + c_{i3}\cos(\pi t/6) \qquad (1.1)$$

should do rather nicely for these data, where $\texttt{Temp}_i$ is the temperature function for the ith weather station, and (c_{i1}, c_{i2}, c_{i3}) is a vector of three parameters associated with that station.

In fact, there are clear departures from sinusoidal or simple harmonic behavior. One way to see this is to compute the function

$$L\texttt{Temp} = (\pi/6)^2 D\texttt{Temp} + D^3\texttt{Temp}. \qquad (1.2)$$

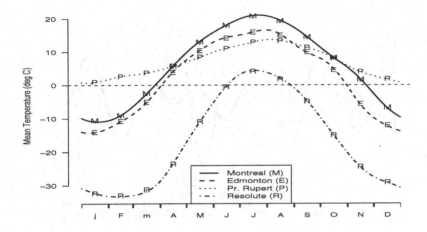

Fig. 1.11 Mean temperatures at four Canadian weather stations.

The notation D^mTemp means "take the mth derivative of function Temp," and the notation LTemp stands for the function which results from applying the *linear differential operator* $L = (\pi/6)^2 D + D^3$ to the function Temp. The resulting function, LTemp, is often called a *forcing function*. If a temperature function is truly sinusoidal, then LTemp should be exactly zero, as it would be for any function of the form (1.1). That is, it would conform to the *differential equation*

$$L\text{Temp} = 0 \quad \text{or} \quad D^3\text{Temp} = -(\pi/6)^2 D\text{Temp}.$$

But Figure 1.12 indicates that the functions LTemp$_i$ display systematic features that are especially strong in the summer and autumn months. Put another way, temperature at a particular weather station can be described as the solution of the *nonhomogeneous* differential equation corresponding to LTemp $= u$, where the forcing function u can be viewed as input from outside of the system, or as an exogenous influence. Meteorologists suggest, for example, that these spring and autumn effects are partly due to the change in the reflectance of land when snow or ice melts, and this would be consistent with the fact that the least sinusoidal records are associated with continental stations well separated from large bodies of water.

Here, the point is that we may often find it interesting to remove effects of a simple character by applying a differential operator, rather than by simply subtracting them. This exploits the intrinsic smoothness in the process. Long experience in the natural and engineering sciences suggests that this may get closer to the underlying driving forces at work than just adding and subtracting effects, as is routinely done in multivariate data analysis. We will consider this idea in depth in Chapter 11.

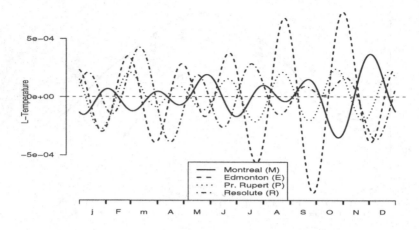

Fig. 1.12 The result of applying the differential operator $L = (\pi/6)^2 D + D^3$ to the estimated temperature functions in Figure 1.11. If the variation in temperature were purely sinusoidal, these curves would be exactly zero.

1.5 First Steps in a Functional Data Analysis

1.5.1 Data Representation: Smoothing and Interpolation

Assuming that a functional datum for replication i arrives as a finite set of measured values, $y_{i1}, \ldots, y_{in}$, the first task is to convert these values to a function x_i with values $x_i(t)$ computable for any desired argument value t. If these observations are assumed to be errorless, then the process is *interpolation*, but if they have some observational error that needs removing, then the conversion from (finite) data to functions (which can theoretically be evaluated at an infinite number of points) may involve *smoothing*.

Chapter 5 offers a survey of these procedures. The *roughness penalty* smoothing method discussed there will be used much more broadly in many contexts throughout the book, and not merely for the purpose of estimating a function from a set of observed values. The daily precipitation data for Prince Rupert, one of the wettest places on the continent, is shown in Figure 1.13. The curve in the figure, which seems to capture the smooth variation in precipitation, was estimated by penalizing the squared deviations in *harmonic acceleration* as measured by the differential operator (1.2).

The gait data in Figure 1.6 were converted to functions by the simplest of interpolation schemes: joining each pair of adjacent observations by a straight line segment. This approach would be inadequate if we required derivative information. However,

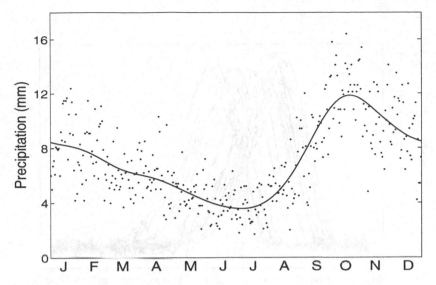

Fig. 1.13 The points indicate average daily rainfall at Prince Rupert on the northern coast of British Columbia. The curve was fit to these data using a roughness penalty method.

one might perform a certain amount of smoothing while still respecting the periodicity of the data by fitting a Fourier series to each record: A constant plus three pairs of sine and cosine terms does a reasonable job for these data. The growth data in Figures 1.1, 1.2, and 1.15 were fit using smoothing splines. The temperature data in Figure 1.11 were fit smoothing a finite Fourier series. This more sophisticated technique can also provide high-quality derivative information.

There are often conceptual constraints on the functions that we estimate. For example, a smooth of precipitation such as that in Figure 1.13 should logically never be negative. There is no danger of this happening for a station as moist as Prince Rupert, but a smooth of the data in Resolute, the driest place that we have data for, can easily violate this constraint. The growth curve fits should be strictly increasing, and we shall see that imposing this constraint results in a rather better estimate of the acceleration curves that we saw in Figure 1.2. Chapter 5 shows how to fit a variety of constrained functions to data.

1.5.2 Data Registration or Feature Alignment

Figure 1.14 shows some biomechanical data. The curves in the figure are 20 records of the force exerted on a meter during a brief pinch by the thumb and forefinger. The subject was required to maintain a certain background force on a force meter and then to squeeze the meter aiming at a specified maximum value, returning af-

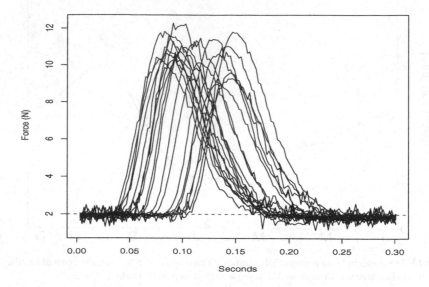

Fig. 1.14 Twenty recordings of the force exerted by the thumb and forefinger where a constant background force of 2 newtons was maintained prior to a brief impulse targeted to reach 10 newtons. Force was sampled 500 times per second.

terwards to the background level. The purpose of the experiment was to study the neurophysiology of the thumb–forefinger muscle group. The data were collected at the MRC Applied Psychology Unit, Cambridge, by R. Flanagan (Ramsay et al. 1995b).

These data illustrate a common problem in functional data analysis. The start of the pinch is located arbitrarily in time, and a first step is to align the records by some shift of the time axis. In Chapter 8 we take up the question of how to estimate this shift and how to go further if necessary to estimate record-specific linear or nonlinear transformations of the argument.

1.5.3 Graphing Functional Data

Displaying the results of a functional data analysis can be a challenge. With the gait data in Figures 1.6 and 1.7, we have already seen that different displays of data can bring out different features of interest, and the standard plot of $x(t)$ against t is not necessarily the most informative. It is impossible to be prescriptive about the best type of plot for a given set of data or procedure, but we shall give illustrations of various ways of plotting the results. These are intended to stimulate the reader's imagination rather than to lay down rigid rules.

1.5.4 Plotting Pairs of Derivatives: Phase-Plane Plots

Let us look at a couple of plots to explore the possibilities opened up by access to derivatives of functions. Figure 1.15 contains *phase-plane plots* of the female height curves in Figure 1.1, consisting of plots of the accelerations or second derivatives against their velocities or first derivatives. Each curve begins in the lower right in infancy, with strong positive velocity and negative acceleration. The middle of the pubertal growth spurt for each girl corresponds to the point where her velocity is maximized after early childhood. The circles mark the position of each girl at age 11.7, the average midpubertal age. The pubertal growth loop for each girl is entered from the right and below, usually after a cusp or small loop. The acceleration is positive for a while as the velocity increases until the acceleration drops again to zero on the right at the middle of the spurt. The large negative swing terminates near the origin where both velocity and acceleration vanish at the beginning of adulthood.

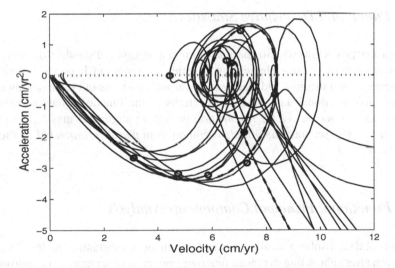

Fig. 1.15 The second derivative or acceleration curves are plotted against the first derivative or velocity curves for the ten female growth curves in Figure 1.1. Each curve begins in time off the lower right with the strong velocity and deceleration of infant growth. The velocities and accelerations at age 11.7 years, the average age of the middle of the growth spurt, are marked on each curve by circles. The curve is highlighted by a heavy dashed line is that of a girl who goes through puberty at the average age.

Many interesting features in this plot demand further consideration. Variability is greatest in the lower right in early childhood, but it is curious that two of the 10 girls have quite distinctive curves in that region. Why does the pubertal growth spurt show up as a loop? What information does the size of the loop convey? Why are the larger loops tending to be on the right and the smaller on the left? We see from the shapes of the loop and from the position of the 11.7 year marker that girls with early pubertal spurts (marker point well to the left) tend to have very large loops, and late-spurt girls have small ones. Does interchild variability correspond to something like growth energy? Clearly there must be a lot of information in how velocity and acceleration are linked together in human growth, and perhaps in many other processes as well.

1.6 Exploring Variability in Functional Data

The examples considered so far offer a glimpse of ways in which the variability of a set of functional data can be interesting, but there is a need for more detailed and sophisticated ways of investigating variability. These are a major theme of this book.

1.6.1 Functional Descriptive Statistics

Any data analysis begins with the basics: estimating means and standard deviations. Functional versions of these elementary statistics are given in Chapter 7. But what is elementary for univariate and classic multivariate data turns out to be not always so simple for functional data. Chapter 8 returns to the functional data summary problem, and shows that *curve registration* or feature alignment may have to be applied in order to separate *amplitude variation* from *phase variation* before these statistics are used.

1.6.2 Functional Principal Components Analysis

Most sets of data display a small number of dominant or substantial modes of variation, even after subtracting the mean function from each observation. An approach to identifying and exploring these, set out in Chapter 7, is to adapt the classic multivariate procedure of principal components analysis to functional data. Techniques of smoothing are incorporated into the functional principal components analysis itself, thereby demonstrating that smoothing methods have a far wider rôle in functional data analysis than merely in the initial step of converting a finite number of observations to functional form.

1.6.3 Functional Canonical Correlation

How do two or more sets of records covary or depend on one another? While studying Figure 1.7, we might consider how correlations embedded in the record-to-record variations in hip and knee angles might be profitably examined and used to further our understanding the biomechanics of walking.

The functional linear modeling framework approaches this question by considering one of the sets of functional observations as a covariate and the other as a response variable. In many cases, however, it does not seem reasonable to impose this kind of asymmetry. We shall develop two rather different methods that treat both sets of variables in an even-handed way. One method essentially treats the pair $(\texttt{Hip}_i, \texttt{Knee}_i)$ as a single vector-valued function, and then extends the functional principal components approach to perform an analysis. Chapter 7 takes another approach, a functional version of canonical correlation analysis, identifying components of variability in each of the two sets of observations which are highly correlated with one another.

For many of the methods we discuss, a naïve approach extending the classic multivariate method will usually give reasonable results, though regularization will often improve these. However, when a linear predictor is based on a functional observation, and also in functional canonical correlation analysis, imposing smoothness on functional regression coefficients is not an optional extra, but rather an intrinsic and necessary part of the analysis; the reasons are discussed in Chapters 7 and 8.

1.7 Functional Linear Models

The techniques of linear regression, analysis of variance, and linear modeling all investigate the way in which variability in observed data can be accounted for by other known or observed variables. They can all be placed within the framework of the linear model

$$y = \mathbf{Z}\beta + \varepsilon \tag{1.3}$$

where, in the simplest case, y is typically a vector of observations, β is a parameter vector, $\mathbf{Z}$ is a matrix that defines a linear transformation from parameter space to observation space, and ε is an error vector with mean zero. The design matrix $\mathbf{Z}$ incorporates observed covariates or independent variables.

To extend these ideas to the functional context, we retain the basic structure (1.3) but allow more general interpretations of the symbols within it. For example, we might ask of the Canadian weather data:

- If each weather station is broadly categorized as being Atlantic, Pacific, Continental or Arctic, in what way does the geographical category characterize the detailed temperature profile `Temp` and account for the different profiles observed? In Chapter 10 we introduce a functional analysis of variance methodology, where both the parameters and the observations become functions, but the matrix $\mathbf{Z}$ remains the same as in the classic multivariate case.

- Could a temperature record Temp be used to predict the logarithm of total annual precipitation? In Chapter 9 we extend the idea of linear regression to the case where the independent variable, or covariate, is a function, but the response variable (log total annual precipitation in this case) is not.
- Can the temperature record Temp be used as a predictor of the entire precipitation profile, not merely the total precipitation? This requires a fully functional linear model, where all the terms in the model have more general form than in the classic case. This topic is considered in Chapter 10.
- We considered earlier the many roles that derivatives play in functional data analysis. In the functional linear model, we may use derivatives as dependent and independent variables. Chapter 10 is a first look at this idea, and sets the stage for the following chapters on differential equations.

1.8 Using Derivatives in Functional Data Analysis

In Section 1.4 we have already had a taste of the ways in which derivatives and linear differential operators are useful in functional data analysis. The use of derivatives is important both in extending the range of simple graphical exploratory methods and in the development of more detailed methodology. This is a theme that will be explored in much more detail in Chapter 11, but some preliminary discussion is appropriate here.

Chapter 11 takes up the question, unique to functional data analysis, of how to use derivative information in studying components of variation. An approach called *principal differential analysis* identifies important variance components by estimating a linear differential operator that will annihilate them (if the model is adequate). Linear differential operators, whether estimated from data or constructed from external modeling considerations, also play an important part in developing regularization methods more general than those in common use.

1.9 Concluding Remarks

In the course of the book, we shall describe a considerable number of techniques and algorithms to explain how the methodology we develop can actually be used in practice. We shall also illustrate this methodology on a variety of data sets drawn from various fields, including where appropriate the examples introduced in this chapter. However, it is not our intention to provide a cookbook for functional data analysis.

In broad terms, our goals are simultaneously more ambitious and more modest: more ambitious by encouraging readers to think about and understand functional data in a new way but more modest in that the methods described in this book are hardly the last word in how to approach any particular problems. We believe that readers will gain more by experimenting with various modifications of the principles described herein than by following any suggestion to the letter. To make this easier, script files like "fdarm-ch01.R" in the "scripts" subdirectory of the companion "fda" package for R can be copied and revised to test understanding of the concepts. The "debug" function in R allows a user to walk through standard R code line by line with real examples until any desired level of understanding is achieved.

For those who would like access to the software that we have used, a selection is available on the website:

$$\texttt{http://www.functionaldata.org}$$

and in the `fda` package in R. This website will also be used to publicize related and future work by the authors and others, and to make available the data sets referred to in the book that we are permitted to release publicly.

1.10 Some Things to Try

In this and subsequent chapters, we suggest some simple exercises that you might consider trying.

1. Find some samples of functional data and plot them. Make a short list of questions that you have about the processes generating the data. If you do not have some data laying around in a file somewhere, here are some suggestions:

   a. Use your credit card or debit/bank card transactions in your last statement. If you keep your statements or maintain an electronic record, consider entering also the statements for five or so previous months or even for the same month last year.

   b. Bend over and try to touch your toes. Please do not strain! Have someone measure how far your fingers are from the floor (or your wrist if you are that flexible). Now inhale and exhale slowly. Remeasure and repeat for a series of breath cycles. Now repeat the exercise, but for the person doing the measuring.

   c. Visit some woods and count the number of birds that you see or the number of varieties. Do this for a series of visits, spread over a day or a week. If over a week, record the temperature and cloud and precipitation status as well.

   d. Visit a weather website, and record the five-day temperature forecasts for a number of cities.

   e. If you have a chronic concern like allergies, brainstorm a list of terms to describe the severity of the condition, sort the terms from mild to severe and assign numbers to them. Also brainstorm a list of possible contributing factors and develop a scale for translating variations in each contributing factor into numbers. Each day record the level of the condition and each potential contributing factor. One of us solved a serious allergy problem doing this.

Chapter 2
Essential Comparisons of the Matlab and R Languages

We assume a working knowledge of either Matlab or R. For either language, there are many books that describe the basics for beginners. However, a brief comparison of the two languages might help someone familiar with one language read code written in the other.

Matlab and R have many features in common. Some of the differences are trivial while others can be troublesome. Where differences are minor, we offer code in only one language, which will be often R.

We will use typewriter font for any text meant to be interpreted as Matlab or R code, such as `plot(heightfd)`.

2.1 A Quick Comparison of Matlab and R Syntax

There are similarities and differences in the syntax for Matlab and R.

2.1.1 Minor Differences

Here is a quick list of the more commonly occurring differences so that you easily translate a command in one language in that in the other:

- Your code will be easier to read if function names describe what the function does. This often produces a preference for names with words strung together. This is often done in Matlab by connecting words or character strings with underscores like `create_fourier_basis`. This is also acceptable in R. However, it is not used that often, because previous versions of R (and S-Plus) accepted an underscore as a replacement operator. Names in R are more likely to use dots or periods to separate strings, as in `create.fourier.basis` used below.

J.O. Ramsay et al., *Functional Data Analysis with R and MATLAB*, Use R,
DOI: 10.1007/978-0-387-98185-7_2,

- The dot or period in Matlab identifies a component of a `struct array`. This is roughly comparable to the use of the dollar sign ($) in R to identify components of a list, though there are differences, which we will not discuss here.
- Vectors are often defined using the `c()` command in R, as in `rng = c(0,1)`. In Matlab, this is accomplished using square brackets, as in `rng = [0,1]`.
- On the other hand, R uses square brackets to select subsets of values from a vector, such as `rng[2]`. Matlab does this with parentheses, as in `rng(2)`.
- R has logical variables with values either `TRUE` or `FALSE`. Recent releases of Matlab also have logical variables taking values `true` or `false`.
- Previous releases of R, S, and S-Plus allowed the use of T and F for `TRUE` and `FALSE`. Recent releases of R have allowed users to assign other values to T or F for compatibility with other languages. This has the unfortunate side effect that R code written using T or F could throw an error or give a wrong answer without warning if, for example, a user defined `F = TRUE` or `F = c('Do', 'not', 'use', 'F', 'as', 'a', 'logical.')`.
- In both languages, numbers can sometimes be used as logicals; in such cases, 0 is treated as FALSE and any nonzero is TRUE.
- If a line of code is not syntactically complete, the R interpreter looks for that code to continue on the next line; Matlab requires the line to end in "..." if the code is to be continued on the next line.
- Matlab normally terminates a command with a semicolon. If this is not done, Matlab automatically displays the object produced by the command. Lines in R can end in a semicolon, but that is optional.

In this book, where we give commands in both languages, the R version will come first and the Matlab version second. But we will often give only one version; in most such cases, the conversion is just a matter of following these rules.

The matter of the *assignment operator* needs at least a brief comment. In R the correct way to write the transfer of the value produced by the right side of a statement to the object named on the left side is with the two-character sequence <-. We like this notation, and prefer to use it in our own work. However, there was from the beginning a resistance among users of R, S and S-PLUS to the use of two characters instead of one. The underscore _ was allowed but created problems, if only because of incompatibility with many other languages like Matlab that allowed the underscore in names. Recent versions of R allow the use of = for replacement in most contexts, but users are warned that there are situations where the code becomes ambiguous and may generate errors that can be hard to trace. With this in mind, we notwithstanding opt for = in this book, primarily to keep statements readable and to minimize the differences between R and Matlab. (Matlab uses only = for replacement.)

2.1.2 Using Functions in the Two Languages

The ways in which arguments are passed to functions and computed results returned
is, unfortunately, different in the two languages. We can illustrate the differences by
the ways in which we use the important smoothing function, smooth.basis in R
and smooth_basis in Matlab. Here is a full function call in R:

```
smoothlist = smooth.basis(argvals, y, fdParobj,
                          wtvec, fdnames)
```

and here is the Matlab counterpart:

```
[fdobj, df, gcv, coef, SSE, penmat, y2cMap] = ...
   smooth_basis(argvals, y, fdParobj, wtvec, fdnames);
```

An R function outputs only a single object, so that if multiple objects need to
be returned, as in this example, then R returns them within a list object. But Matlab
returns its outputs as a set of variable(s); if more than one, their names are contained
within square brackets.

The handy R feature of being able to use argument names to provide any subset
of arguments in any order does not exist in Matlab. Matlab function calls require
the arguments in a rigid order, though only a subsequence of leading arguments can
be supplied. The same is true of the outputs. Consequently, Matlab programmers
position essential arguments and returned objects first.

For example, most of the time we just need three arguments and a single output
for smooth.basis and its Matlab counterpart, so that a simpler R call might be

```
myfdobj = smooth.basis(argvals, y, fdParobj)$fd
```

and the Matlab version would be

```
myfdobj = smooth_basis(argvals, y, fdParobj);
```

Here R gets around the fact that it can only return a single object by returning a
list and using the $fd suffix to select from that list the object required. Matlab just
returns the single object. If we want the third output gcv, we could get that in R by
replacing fd with gcv; in Matlab, we need to provide explicit names for undesired
outputs as, [fdobj, df, gcv] in this example. R also has the advantage of
being able to change the order of arguments by a call like

```
myfdobj = smooth.basis(y=yvec, argvals=tvec,
                       fdParobj)$fd
```

In order to keep things simple, we will try keep the function calls as similar as
possible in the examples in this book.

2.2 Singleton Index Issues

The default behavior in matrices and arrays with a singleton dimension is exactly the opposite between R and Matlab: R drops apparently redundant dimensions, compressing a matrix to a vector or an array to a matrix or vector. Matlab does not.

For example, `temp = matrix(c(1,2,3,4),2,2)` sets up a 2×2 matrix in R, and `class(temp)` tells us this is a `"matrix"`. However, `class(temp[,1])` yields `"numeric"`, which says that `temp[,1]` is no longer a matrix. If you want a matrix from this operation, use `temp[,1, drop=FALSE]`. This can have unfortunate consequences in that an operation that expects `temp[,index]` to be a matrix will work when `length(index) > 1` but may throw an error when `length(index) = 1`. If A is a three-dimensional array, `A1 = A[,1,]` will be a matrix provided the first and third dimensions of A both have multiple levels. If this is in doubt, `dim(A1) = dim(A)[-2]` will ensure that A is a matrix, not a vector as it would be if the first or third dimensions of A were singleton.

Matlab has the complementary problem. An array with a single index, as in `temp = myarray(:,1,:)`, is still an array with the same number of dimensions. If you want to multiply this by a matrix or plot its columns, the `squeeze()` function will eliminate unwanted singleton dimensions. In other words, `squeeze(temp)` is a matrix, as long as only one of the three dimension of `temp` is a singleton.

A user who does not understand these issues in R or Matlab can lose much time programming around problems that are otherwise easily handled.

2.3 Classes and Objects in R and Matlab

Our code uses *object-oriented programming*, which brings great simplicity to the use of some of the functions. For example, we can use the *plot* command in either language to create specialized graphics tailored to the type of object being plotted, e.g., for basis function systems or functional data objects, as we shall see in the next chapter.

The notion of a class is built on the more primitive notion of a *list* object in R and its counterpart in Matlab, a *struct* object. Lists and structs are used to group together types of information with different internal characteristics. For example, we might want to combine a vector of numbers with a fairly lengthy name or string that can be used as a title for plots. The vector of numbers is a *numeric* object in R or a *double* object in Matlab, while the title string is a *character* object in R and a *char* object in Matlab.

Once we have this capacity of grouping together things with arbitrary properties, it is an easy additional step to define a class as a specific recipe or predefined combination of types of information, along with a name to identify the name of the recipe. For example, in the next chapter we will define the all-important class `fd` as, minimally, a coefficient array combined with a recipe for a set of basis functions.

That is, an `fd` object is either a `list` or a `struct`, depending on the language, which contains at least two pieces of information, each prespecified and associated with the class name `fd`. Actually, the specification for the basis functions is itself also a object of a specific class, the `basisfd` class in R and the `basis` class in Matlab, but let us save these details until the next chapter.

Unfortunately, the languages differ dramatically in how they define classes, and this has wide-ranging implications. In Matlab, a class is set up as a folder or directory of functions used to work with objects of that class.

R has two different ways to define classes and operations on objects of different classes, the `S3` and `S4` systems. The `fda` package for R uses the `S3` system. In this `S3` standard, R recognizes an object to be of a certain class, e.g., `fd`, solely by the possession of a `'class'` attribute with value `'fd'`. The `class` attribute is used by *generic* functions such as `plot` by *methods dispatch*, which looks first for a function with the name of the generic followed by the class name separated by a period, e.g., `plot.fd` to plot an object of class `fd`.

An essential operation is the extraction of information from an object of a particular class. Each language has simple classes that are basic to its structure, such as the class `matrix` in either language. However, the power of an object-oriented program becomes apparent when a programmer sets up new classes of objects that, typically, contain multiple entities or components that may be named. These components are themselves objects of various classes, which may be among those that are basic to the language or in turn are programmer–constructed new classes.

There are two standards in R for "object oriented" programming, called "S3" and "S4". The `fda` package for R uses the S3 system, which is described in Appendix A of Chambers and Hastie (1991). (The S4 system is described in Chambers (2008).) In the S3 system, everything is a vector. Basic objects might be vectors of numbers, either double precision or integers. Or they might be a vector of character strings of varying length. This differs from Matlab, where a character vector is a vector of single characters; to store names with multiple characters, you must use either a character matrix (if all names have the same number of characters) or a cell array of strings (to store names of different lengths). A list is a vector of pointers to other objects. In R, component `i` of vector `v` is accessed as `v[i]`, except that if `v` is a list, `v[i]` will be a sublist of `v`, and `v[[i]]` will return that component. Objects in R can also have `attributes`, and if an object has a `class` attribute, then it is an object of that class. (Classes in the S4 system are much more rigidly defined; see Chambers (2008).) If a list `xxx` in R has an attribute names = "a", "b", "c", say, these attributes can optionally be accessed via `xxx$a`, `x$b`, `x$c`. To see the names associated with an object x, use `names(x)` or `attr(x, 'names')`. To see all the `attributes` of an object x, use `attributes(x)`. To get a compact summary of x, use `str(x)`. Or the class name can also be used in the language's help command: `help myclass` and `doc myclass` in Matlab or `?myclass` and `help(myclass)` in R.

For example, there are many reasons why one would want to get the coefficient array contained in the functional data `fd` class. In Matlab we do this by using functions that usually begin with the string `get`, as in the command `coefmat =`

getcoef(fdobj) that extracts the coefficient array from object fdobj of the fd class. Similarly, a coefficient array can be inserted into a Matlab fd object with the command fdobj = putcoef(fdobj, coefmat). In Matlab, all the extraction functions associated with a class can be accessed by the command methods myclass.

The procedure for extracting coefficients from an R object depends on the class of the object. If obj is an object of class fd, fdPar or fdSmooth, coef(obj) will return the desired coefficients. (The fd class is discussed in Chapter 4, and the fdPar and fdSmooth classes are discussed in Chapter 5.) This is quite useful, because without this generic function, a user must know more about the internal structure of the object to get the desired coefficients. If obj has class fd, then obj$coefs is equivalent to coef(obj). However, if obj is of class fdPar or fdSmooth, then obj$coefs will return NULL; objfdcoefs will return the desired coefficients.

As of this writing, a "method" has not yet been written for the generic coef function for objects of class monfd, returned by the smooth.monotone function discussed in Section 5.4.2. If obj has that class, it is not clear what a user might want, because it has two different types of coefficients: obj$Wfdobj$coefs give the coefficients of a functional data object that is exponentiated to produce something that is always positive and integrated to produce a nondecreasing function. This is then shifted and scaled by other coefficients in obj$beta to produce the desired monotonic function. In this case, the structure of objects of class monfd is described in the help page for the smooth.monotone function. However, we can also get this information using str(obj), which will work for many other objects regardless of the availability of a suitable help page.

To find the classes for which methods have been written for a particular generic function like coef, use methods('coef'). Conversely, to find generic functions for which methods of a particular class have been written, use, e.g., methods(class='fd'). Unfortunately, neither of these approaches is guaranteed to find everything, in part because of "inheritance' of classes, which is beyond the scope of the present discussion. For more on methods in R, see Appendix A in Chambers and Hastie (1991).

2.4 More to Read

For a more detailed comparison of R and Matlab, see Hiebeler (2009).

There is by now a large and growing literature on R, including many documents of various lengths freely downloadable from the R website:

    http://www.r-project.org

This includes books with brief reviews and publisher information as well as freely downloadable documents in a dozen different languages from Chinese to Vietnamese via "Documents: Other" both on the main R page and from CRAN. This is in addition to documentation beyond `help` that comes with the standard R installation available from `help.start()`, which opens a browser with additional documentation on the language, managing installation, and "Writing R Extensions."

Chapter 3
How to Specify Basis Systems for Building Functions

We build functions in two stages:

1. First, we define a set of functional building blocks ϕ_k called *basis functions*.
2. Then we set up a vector, matrix, or array of coefficients to define the function as a linear combination of these basis functions.

This chapter is primarily about setting up a basis system. The next chapter will discuss the second step of bundling a set of coefficient values with the chosen basis system.

The functions that we wish to model tend to fall into two main categories: periodic and nonperiodic. The Fourier basis system is the usual choice for periodic functions, and the spline basis system (and bsplines in particular) tends to serve well for nonperiodic functions. We go into these two systems in some detail, and the spline basis especially requires considerable discussion. These two systems are often supplemented by the constant and monomial basis systems, and other systems are described more briefly.

A set of functions in both languages are presented for displaying, evaluating and plotting basis systems as well as for other common tasks.

3.1 Basis Function Systems for Constructing Functions

We need to work with functions with features that may be both unpredictable and complicated. Consequently, we require a strategy for constructing functions that works with parameters that are easy to estimate and that can accommodate nearly any curve feature, no matter how localized. On the other hand, we do not want to use more parameters than we need, since doing so would greatly increase computation time and complicate our analyses in many other ways as well.

We use a set of functional building blocks $\phi_k, k = 1, \ldots, K$ called *basis functions*, which are combined linearly. That is, a function $x(t)$ defined in this way is expressed in mathematical notation as

J.O. Ramsay et al., *Functional Data Analysis with R and MATLAB*, Use R,
DOI: 10.1007/978-0-387-98185-7_3,
© Springer Science + Business Media, LLC 2009

$$x(t) = \sum_{k=1}^{K} c_k \phi_k(t) = \mathbf{c}'\phi(t), \tag{3.1}$$

and called a *basis function expansion*. The parameters $c_1, c_2, \ldots, c_K$ are the *coefficients* of the expansion. The matrix expression in the last term of (3.1) uses $\mathbf{c}$ to stand for the vector of K coefficients and ϕ to denote a vector of length K containing the basis functions.

We often want to consider a sample of N functions, $x_i(t) = \sum_{k=1}^{K} c_{ik} \phi_k(t), i = 1, \ldots, N$, and in this case matrix notation for (3.1) becomes

$$\mathbf{x}(t) = \mathbf{C}\phi(t), \tag{3.2}$$

where $\mathbf{x}(t)$ is a vector of length N containing the functions $x_i(t)$, and the coefficient matrix $\mathbf{C}$ has N rows K columns.

Two brief asides on notation are in order here. We often need to distinguish between referring to a function in a general sense and referring to its value at a specific argument value t. Expression (3.1) refers to the basis function expansions of the value of function x at argument value t, but the expansion of x is better written as

$$x = \sum_{k=1}^{K} c_k \phi_k = \mathbf{c}'\phi. \tag{3.3}$$

We will want to indicate the result of taking the mth derivative of a function x, and we will often refer to the first derivative, $m = 1$, as the *velocity* of x and to the second derivative, $m = 2$, as its *acceleration*. No doubt readers will be familiar with the notation

$$\frac{dx}{dt}, \frac{d^2x}{dt^2}, \ldots, \frac{d^m x}{dt^m}$$

used in introductory calculus courses. In order to avoid using ratios in text, and for a number of other reasons, we rather prefer the notation Dx and D^2x for the velocity and acceleration of x, and so on. The notation can also be extended to zero and negative values of m, since $D^0x = x$ and $D^{-1}x$ refers to the indefinite integral of x from some unspecified origin.

The notion of a basis system is hardly new; a polynomial such as $x(t) = 18t^4 - 2t^3 + \sqrt{17}t^2 + \pi/2$ is just such a linear combination of the *monomial* basis functions $1, t, t^2, t^3$, and t^4 with coefficients $\pi/2, 0, \sqrt{17}, -2$, and 18, respectively. Within the monomial basis system, the single basis function 1 is often needed by itself, and we call it the *constant* basis system.

But polynomials are of limited usefulness when complex functional shapes are required. Therefore we do most of our heavy lifting with two basis systems: *splines* and *Fourier series*. These two systems often need to be supplemented by the *constant* and *monomial* basis systems. These four systems can deal with most of the applied problems that we are see in practice.

For each basis system we need a function in either R or Matlab to define a specific set of K basis functions ϕ_k's. These are the `create` functions. Here are the calling statements of the `create` functions in R that set up constant, monomial, Fourier and spline basis systems, omitting arguments that tend only to be used now and then as well as default values:

```
basisobj = create.constant.basis(rangeval)
basisobj = create.monomial.basis(rangeval, nbasis)
basisobj = create.fourier.basis(rangeval, nbasis,
                                        period)
basisobj = create.bspline.basis(rangeval, nbasis,
                                    norder, breaks)
```

We will take each of these functions up in detail below, where we will explain the roles of the arguments. The Matlab counterparts of these create functions are:

```
basisobj = create_constant_basis(rangeval);
basisobj = create_monomial_basis(rangeval, nbasis);
basisobj = create_fourier_basis(rangeval, nbasis, ...
                                        period);
basisobj = create_bspline_basis(rangeval, nbasis, ...
                                    norder, breaks);
```

In either language, the specific basis system that we set up, named in these commands as basisobj, is said to be a *functional basis object* with the class name basis (Matlab) or basisfd(R). Fortunately, users rarely need to worry about the difference in class name between Matlab and R, as they rarely need to specify the class name directly in either language.

However, we see that the first argument rangeval is required in each create function. This argument specifies the lower and upper limits of the values of argument t and is a vector object of length 2. For example, if we need to define a basis over the unit interval $[0, 1]$, we would use a statement like rangeval = c(0,1) in R or rangeval = [0,1] in Matlab.

The second argument nbasis specifies the number K of basis functions. It does not appear in the constant basis call because it is automatically 1.

Either language can use the class name associated with the object to select the right kind of function for operations such as plotting or to check that the object is appropriate for the task at hand. You will see many examples of this in the examples that we provide.

We will defer a more detailed discussion of the structure of the basis or basisfd class to the end of this chapter since this information will only tend to matter in relatively advanced uses of either language, and we will not, ourselves, use this information in our examples.

We will now look at the nature of each basis system in turn, beginning with the only mildly complicated Fourier basis. Then we will discuss the more challenging B-spline basis. That will be followed by more limited remarks on constant and monomial bases. Finally, we will mention only briefly a few other basis systems that are occasionally useful.

3.2 Fourier Series for Periodic Data and Functions

Many functions are required to repeat themselves over a certain period T, as would
be required for expressing seasonal trend in a long time series. The Fourier series is

$$\phi_1(t) = 1$$
$$\phi_2(t) = \sin(\omega t)$$
$$\phi_3(t) = \cos(\omega t)$$
$$\phi_4(t) = \sin(2\omega t)$$
$$\phi_5(t) = \cos(2\omega t)$$
$$\vdots$$

$$(3.4)$$

where the constant ω is related to the period T by the relation

$$\omega = 2\pi/T.$$

We see that, after the first constant basis function, Fourier basis functions are ar-
ranged in successive sine/cosine pairs, with both arguments within any pair being
multiplied by one of the integers $1, 2, \ldots$ up to some upper limit m. If the series
contains both elements of each pair, as is usual, the number of basis functions is
$K = 1 + 2m$. Because of how we define ω, each basis function repeats itself after T
time units have elapsed.

Only two pieces of information are required to define a Fourier basis system:

- the number of basis functions K and
- the period T,

but the second value T can often default to the range of t values spanned by the
data. We will use a Fourier basis in the next chapter to smooth daily temperature
data. The following commands set up a Fourier basis with $K = 65$ basis functions
in R and Matlab with a period of 365 days:

```
daybasis65 = create.fourier.basis(c(0,365), 65)
daybasis65 = create_fourier_basis([0,365],  65);
```

Note that these function calls use the default of $T = 365$, but if we wanted to specify
some other period T, we would use

```
create.fourier.basis(c(0,365), 65, T)
```

in R.

In either language, if K is even, the create functions for Fourier series add on
the missing cosine and set $K = K + 1$. When this leads to more basis functions than
values to be fit, the code takes steps to avoid singularity problems.

There are situations where periodic functions are defined in terms of only sines or
only cosines. For example, a pure sine series will define functions that have the value
0 at the boundary values 0 and T, while a pure cosine series will define functions

with zero derivatives at these points. Bases of this nature can be set up by selecting only the appropriate terms in the series by either subscripting the basis object or by using a component of the class called `dropind` that contains a vector of indices of basis functions to remove from the final series. For example, if we wanted to set up a Fourier basis for functions centered on zero, we would want to not include the initial constant term, and this could be achieved by either a command like

```
zerobasis = create.fourier.basis(rangeval, nbasis,
                                  dropind=1)
```

or, using a basis object that has already been created, by something like

```
zerobasis = daybasis65[2:65]
```

Here is the complete calling sequence in R for the `create.fourier.basis` in R:

```
create.fourier.basis(rangeval=c(0, 1), nbasis=3,
    period=diff(rangeval), dropind=NULL, quadvals=NULL,
    values=NULL, basisvalues=NULL, names=NULL,
    axes=NULL)
```

A detailed description of the use of the function can be obtained by the command

```
help(create.fourier.basis)  or ?create.fourier.basis
help create_fourier_basis   or doc create_fourier_basis
```

in R and Matlab, respectively.

3.3 Spline Series for Nonperiodic Data and Functions

Splines are piecewise polynomials. Spline bases are more flexible and therefore more complicated than finite Fourier series. They are defined by the range of validity, the knots, and the order. There are many different kinds of splines. In this section, we consider only B-splines.

3.3.1 Break Points and Knots

Splines are constructed by dividing the interval of observation into subintervals, with boundaries at points called *break points* or simply *breaks*. Over any subinterval, the spline function is a polynomial of fixed degree or order, but the nature of the polynomial changes as one passes into the next subinterval. We use the term *degree* to refer the highest power in the polynomial. The *order* of a polynomial is one higher than its degree. For example, a straight line is defined by a polynomial of degree one since its highest power is one, but is of order two because it also has a constant term.

We will assume in this book that the order of the polynomial segments is the same for each subinterval.

A spline basis is actually defined in terms of a set of *knots*. These are related to the break points in the sense that every knot has the same value as a break point, but there may be multiple knots at certain break points.

At each break point, neighboring polynomials are constrained to have a certain number of matching derivatives. The number of derivatives that must match is determined by the number of knots positioned at that break point. If only one knot is positioned at a break point, the number of matching derivatives (including the function value itself) is two less than its order, which ensures that for splines of more than order two the join will be seen to be smooth. This is because a function composed of straight line segments of order two will have only the function value (the derivative or order 0) matching, so the function is continuous but its slope is not; this means that the joins would not be seen as smooth by most standards.

3.3.2 Order and Degree

Order four splines are often used, consisting of cubic polynomial segments (degree three), and the single knot per break point makes the function values and first and second derivative values match.

By default, and in the large majority of applications, there will be only a single knot at every break point except for the boundary values at each end of the whole range of t. The end points, however, are assigned as many knots as the order of the spline, implying that the function value will, typically, drop to zero outside of the interval over which the function is defined.

3.3.3 Examples

Perhaps a couple of simple illustrations are in order. First, suppose we define a function over [0,1] with a single interior break point at, say, 0.5. The cubic spline basis set up in the simplest and most usual way has knots $(0,0,0,0,0.5,1,1,1,1)$ because a cubic spline has order four (degree three), so the end knots appear four times each. Similarly, a linear spline has order two, so a single interior break point at 0.5 translates into knots $(0,0,0.5,1,1)$.

Now, suppose that we want an order two polygonal line, but we want to allow the function value to change abruptly at 0.5. This would be achieved by a knot sequence $(0,0,0.5,0.5,1,1)$. Alternatively, suppose we want to work with cubic splines, but we want to allow the first derivative to change abruptly at 0.5 while the function remains continuous. The knot sequence that does this has three knots placed at 0.5. An illustration of such a situation can be seen in the oil refinery Tray 47 function in Figure 1.5.

You will not have to worry about those multiple knots at the end points; the code takes care of this automatically. You will be typically constructing spline functions where you will only have to supply break points, and if these break points are equally spaced, you will not even have to supply these.

To summarize, spline basis systems are defined by the following:

- the break points defining subintervals,
- the degree or order of the polynomial segments, and
- the sequence of knots.

The number K of basis functions in a spline basis system is determined by the relation

$$number\ of\ basis\ functions\ =\ order\ +\ number\ of\ interior\ knots. \qquad (3.5)$$

By *interior* here we mean only knots that are placed at break points which are not either at the beginning or end of the domain of definition of the function. In the knot sequence examples above, that would mean only knots positioned at 0.5.

3.3.4 B-Splines

Within this framework, however, there are several different basis systems for constructing spline functions. We use the most popular, namely the *B-spline basis system*. Other possibilities are M-splines, I-splines, and truncated power functions. For a more extensive discussion of splines, see, e.g, de Boor (2001) or Schumaker (1981).

Figure 3.1 shows the 13 order four B-splines corresponding to nine equally spaced interior knots over the interval $[0, 10]$, constructed in R by the command

```
splinebasis = create.bspline.basis(c(0,10), 13)
```

or, as we indicated in Chapter 2, by the Matlab command

```
splinebasis = create_bspline_basis([0,10], 13);
```

Figure 3.1 results from executing the command `plot(splinebasis)`.

Aside from the two end basis function, each basis function begins at zero and, at a certain knot location, rises to a peak before falling back to zero and remaining there until the right boundary. The first and last basis functions rise from the first and last interior knot to a value of one on the right and left boundary, respectively, but are otherwise zero. Basis functions in the center are positive only over four intervals, but the second and third basis functions, along with their counterparts on the right, are positive over two and three intervals, respectively. That is, all B-spline basis functions are positive over at most four adjacent intervals. This *compact support* property is important for computational efficiency since the effort required is proportional to K as a consequence, rather than to K^2 for basis functions not having this property.

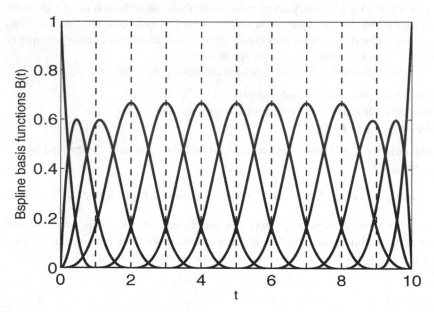

Fig. 3.1 The 13 spline basis functions defined over the interval [0,10] by nine interior boundaries or knots. The polynomial segments are cubic or order four polynomials, and at each knot the polynomial values and their first two derivatives are required to match.

The role of the order of a spline is illustrated in Figure 3.2, where we have plotted linear combinations of spline basis functions of orders two, three and four, called *spline functions*, that best fit a sine function and its first derivative. The three R commands that set up these basis systems are

```
basis2 = create.bspline.basis(c(0,2*pi), 5, 2)
basis3 = create.bspline.basis(c(0,2*pi), 6, 3)
basis4 = create.bspline.basis(c(0,2*pi), 7, 4)
```

Recall from relation (3.5) that, using three interior knots in each case, we increase the number of basis functions each time that we increase the order of the spline basis.

We see in the upper left panel the order two spline function, a polygon, that best fits the sine function, and we see how poorly its derivative, a step function, fits the sine's derivative in the left panel. As we increase order, going down the panels, we see that the fit to both the sine and its derivative improves, as well as the smoothness of these two fits. In general, if we need smooth and accurate derivatives, we need to increase the order of the spline. A useful rule to remember is to *fix the order of the spline basis to be at least two higher than the highest order derivative to be used*. By this rule, a cubic spline basis is a good choice as long as you do not need to look at any of its derivatives.

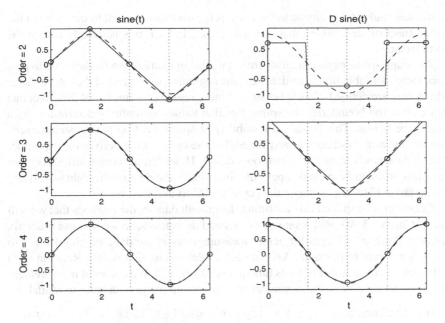

Fig. 3.2 In the left panels, the solid line indicates the spline function of a particular order that fits the sine function shown as a dashed line. In the right panels, the corresponding fits to its derivative, a cosine function, are shown. The vertical dotted lines are the interior knots defining the splines.

The order of a spline is four by default, corresponding to cubic polynomial segments, but if we wanted a basis system with the same knot locations but of order six, we would use an additional argument, as in

```
splinebasis = create.bspline.basis(c(0,10), 15, 6)
```

If, in addition, we wanted to specify the knot locations to be something other than equally spaced, we would use a fourth argument in the function call, with a command such as `create.bspline.basis(c(0,10), nbasis, norder, knotvec)`.

Notice in Figure 3.1 that any single spline basis function is nonzero only over a limited number of intervals, a feature that can be seen more clearly if you use the command `plot(splinebasis[7])` (in R) to plot only the seventh basis function. You would then see that an order four spline basis function is nonzero over four subintervals; similarly, an order six spline is nonzero over over six subintervals. Change 7 to 1 or 2 in this `plot` command to reveal that the end splines are nonzero over a smaller number of intervals.

The B-spline basis system has a property that is often useful: *the sum of the B-spline basis function values at any point t is equal to one.* Note, for example, in Figure 3.1 that the first and last basis functions are exactly one at the boundaries. This is because all the other basis functions go to zero at these end points. Also, because each basis function peaks at a single point, it follows that the value of a coefficient multiplying any basis function is approximately equal to the value of the spline function near where that function peaks. Indeed, this is exactly true at the boundaries.

Although spline basis functions are wonderful in many respects, they tend to produce rather unstable fits to the data near the beginning or the end of the interval over which they are defined. This is because in these regions we run out of data to define them, so at the boundaries the spline function values are entirely determined by a single coefficient. This boundary instability of spline fits becomes especially serious for derivative estimation, and the higher the order of the derivative, the wilder its behavior tends to be at the two boundaries. However, a Fourier series does not have this problem because it is periodic; in essence, the data on the right effectively "wrap around" to help estimate the curve at the left and vice versa.

Let us set up a spline basis for fitting the growth data by the methods that we will use in Chapter 5. We will want smooth second derivatives, so we will use order six splines. There are 31 ages for height measurements in the data, ranging from 1 to 18, and we want to position a knot at each of these sampling points. Relation (3.5) indicates that the number of basis function is $29 + 6 = 35$. If the ages of measurement are in vector age, then the command that will set up the growth basis in Matlab is

```
heightbasis = create_bspline_basis([1,18], 35,6,age);
```

As with the Fourier basis, we can select subsets of B-spline basis functions to define a basis by either dropping basis functions using the dropind argument or by selecting those basis functions that we want by using subscripts.

Here is the complete calling sequence in R for the create.bspline.basis in R:

```
create.bspline.basis(rangeval=NULL, nbasis=NULL,
    norder=4, breaks=NULL, dropind=NULL, quadvals=NULL,
    values=NULL, basisvalues=NULL,
    names="bspl", axes=NULL)
```

A detailed description of the use of the function can be obtained by the commands

```
help(create.bspline.basis)
help create_bspline_basis
```

in R and Matlab, respectively.

3.3.5 Computational Issues Concerning the Range of t

We conclude this section with a tip than can be important if you are using large numbers of spline basis functions. As with any calculation on a computer where the accuracy of results is limited by the number of bits used to express a value, some accuracy can be lost along the way. This can occasionally become serious. A spline is constructed by computing a series of differences. These are especially prone to rounding errors when the values being differenced are close together. To avoid this, you may need to redefine t so that the length of each subinterval is roughly equal to one. For the gait data example shown in Figure 1.6, where we would construct 23 basis functions if we placed a knot at each time of observation, it would be better, in fact, to run time from 0 to 20 than from 0 to 1 as shown. The handwriting example is even more critical, and by changing the time unit from seconds to milliseconds, we can avoid a substantial amount of rounding error.

On the other hand, computations involving Fourier basis functions tend to be more accurate and stable if the interval $[0, T]$ is not too different from $[0, 2\pi]$. We have encountered computational issues, for example, in analyses of the weather data when we worked with $[0, 365]$. Once results have been obtained, it is usually a simple matter to rescale them for plotting purposes to a more natural interval.

3.4 Constant, Monomial and Other Bases

3.4.1 The Constant Basis

Different situations call for different basis systems. One such case leads to the simplest basis system. This is the *constant basis*, which contains only a single function whose value is equal to one no matter what value of t is involved. We need the constant basis surprisingly often. For example, we will see in functional regression and elsewhere that we might need to compare an analysis using an unconstrained time-varying function (represented by a functional data or functional parameter object discussed in Chapters 4 and 5, respectively) with a comparable analysis using a constant. We can also convert a conventional scalar variable into functional form by using the values of that variable as coefficients multiplying the constant basis.

The constant basis over, say $[0, 1]$, is constructed in R by

```
conbasis = create.constant.basis(c(0,1))
conbasis = create_constant_basis([0,1]);
```

in R and Matlab, respectively.

3.4.2 The Monomial Basis

Simple trends in data are often fit by straight lines, quadratic polynomials, and so on. *Polynomial regression* is a topic found in most texts on the linear model or regression analysis, and is, along with Fourier analysis, a form of functional data analysis that has been used in statistics for a long time. As with constant functions, these may often serve as benchmark or reference functions against which spline-based functions are compared.

The basis functions in a monomial basis are the successive powers of t : $1, t, t^2, t^3$ and so on. The number of basis functions is one more than the highest power in the sequence. No parameters other than the interval over which the basis is defined are needed. A basis for cubic polynomials is defined over $[0, 1]$ in R by

```
monbasis = create.monomial.basis(c(0,1), 4)
monbasis = create_monomial_basis([0,1], 4);
```

Be warned that beyond `nbasis = 7`, the monomial basis system functions become so highly correlated with each other that near singularity conditions can arise.

3.4.3 Other Basis Systems

Here are other basis systems available at the time of writing:

- The *exponential basis*, a set of exponential functions, $\exp(\alpha_k t)$, each with a different rate parameter α_k, and created with function `create.exponential.basis`.
- The *polygonal basis*, defining a function made up of straight line segments, and created with function `create.polygonal.basis`.
- The *power basis*, consisting of a sequence of possibly noninteger powers and even negative powers, of an argument t. These bases are created with the function `create.power.basis`. (Negative powers should be avoided if `rangeval`, the interval of validity of the basis set, includes zero.)

Many other basis systems are possible, but so far have not seemed important enough in functional data analysis to justify writing the code required to include them in the `fda` package. However, a great deal of use in applications is made of bases defined *empirically* by the *principal components analysis* of a sample of curves. Basis functions defined in this way are the most compact possible in the sense of providing the best possible fit for fixed K. If one needs a low-dimensional basis system, this is the way to go. Because principal components basis functions, which we call *harmonics*, are also orthogonal, they are often referred to in various fields as *empirical orthogonal functions* or *"eofs"*. Further details are available in Chapter 7.

3.5 Methods for Functional Basis Objects

Common tasks like `plot` are called `generic functions`, for which `methods` are written for object of different classes; see Section 2.3. In R, to see a list of generic functions available for basis objects, use `methods(class='basisfd')`.

Once a basis object is set up, we would like to use some of these `generic functions` via `methods` written for objects of class `basisfd` in R or `basis` in Matlab. Some of the most commonly used generic functions with methods for functional basis objects are listed here. Others requiring more detailed treatment are discussed later. The R function is shown first and the Matlab version second, separated by a /.

In R, the actual name of the function has the suffix .basisfd, but the function is usually used with its initial generic part only, though you may see some exceptions to this general rule. That is, one types print(basisobj) to display the structure of functional basis object basisobj, even though the actual name of the function is print.basisfd. In Matlab, however, the complete function name is required.

print/display The type, range, number of basis functions, and parameters of the functional basis object are displayed. Function print is used in R and display in Matlab. These are invoked if the object name is typed (without a semicolon in Matlab).

summary A more compact display of the structure of the basis object.

==/eq The equality of two functions is tested, and a logical value returned, as in basis1 == basis2 in R or eq(basis1,basis2) in Matlab.

is/isa_basis Returns a logical value indicating whether the object is a functional basis object. In R the function inherits is similar.

In R, we can extract or insert/replace a component of a basis object, such as its params vector, by using the component name preceded by $, as in basisobj$ params. This is a standard R protocol for accessing components of a list. In Matlab, there is a separate function for each component to be extracted. Not all components of an object can be changed safely; some component values interlock with others to define the object, and if you change these, you may later get a cryptic error message or (worse) erroneous results. But for those less critical components, which include *container components* dropind, quadvals, basisvalues and values, the R procedure is simple. The object name with the $ suffix appears on the left side of the assignment operator. In Matlab, each reasonable replacement operation has its own function, beginning with put. The first argument in the function is the name of the basis object, and the second argument is the object to be extracted or inserted. The names of these extractor and insertion functions are displayed in Table 3.1 in Section 3.6.

It is often handy to set up a matrix of basis function values, say for some specialized plotting operation or as an input into a regression analysis. To this end, we have the basis *evaluation* functions

```
basismatrix = eval.basis(tvec, mybasis)
basismatrix = eval_basis(tvec, mybasis)
```

where argument tvec is a vector of n argument values within the range used to define the basis, and argument mybasis is the name of the basis system that you have created. The resulting basismatrix is n by K. One can also compute the derivatives of the basis functions by adding a third argument that specifies the degree of the derivative, as in

```
Dbasismatrix = eval.basis(tvec, mybasis, 1)
Dbasismatrix = eval_basis(tvec, mybasis, 1)
```

Warning: Do not use the command `eval` without its suffix; this command is a part of the core system in both languages and reserved for something quite different. For example, in R `print(mybasis)` does "methods dispatch," passing `mybasis` to function `print.basisfd`. However, `eval(tvec, mybasis)` does not invoke `eval.basis(mybasis)`.

An alternative in R is the generic `predict` function. With this, the previous two function calls could be accomplished as follows:

```
basismatrix = predict(mybasis, tvec)
Dbasismatrix = predict(mybasis, tvec, 1)
```

In the execution of these two commands, the (S3) "methods dispatch" in R searches for a function with the name of the generic combined with the name of the class of the first argument. In this case this process will find `predict.basisfd`, which in turn is a wrapper for `eval.basis`.

There are `predict` methods written for many different classes of objects, which makes it easier to remember the function call. Moreover, for objects with similar functionality but different structure, a user does not have to know the exact class of the object. We use this later with objects of class `fd` and `fdSmooth`, for example.

3.6 The Structure of the `basisfd` or `basis` Class

All basis objects share a common structure, and all of the `create` functions are designed to make the call to the function `basisfd` in R or `basis` in Matlab more convenient. Functions like these two that set up objects of a specific class are called *constructor* functions. The complete calling sequence for `basisfd` in R is

```
basisfd(type, rangeval, nbasis, params,
    dropind=vector("list", 0),
    quadvals=vector("list", 0),
    values=vector("list", 0),
    basisvalues=vector("list", 0))
```

The equivalent Matlab calling sequence lacks specification of default values:

```
basis(basistype, rangeval, nbasis, params, dropind,
    quadvals, values, basisvalues)
```

We include a brief description of each argument here for R users, but you should use the `help` command in either language to get more information.

type   A character string indicating the type of basis. A number of character sequences are permitted for each type to allow for abbreviations and optional capitalization.

rangeval   A vector of length two containing the lower and upper boundaries of the range over which the basis is defined. If a positive number if supplied instead, the lower limit is set to zero.

nbasis   The number of basis functions.

params   A vector of parameter values defining the basis. If the basis type is "fourier", this is a single number indicating the period. That is, the basis functions are periodic on the interval (0,PARAMS) or any translation of it. If the basis type is bspline, the values are interior knots at which the piecewise polynomials join.

dropind   A vector of integers specifying the basis functions to be dropped, if any. For example, if it is required that a function be zero at the left boundary, this is achieved by dropping the first basis function, the only one that is nonzero at that point.

The final three arguments, quadvals, values, and basisvalues, are used to store basis function values in situations where a basis system is evaluated repeatedly.

quadvals   A matrix with two columns and a number of rows equal to the number of argument values used to approximate an integral (e.g., using Simpson's rule). The first column contains the argument values. A minimum of five values is required. For type = 'bspline', this is used in each interknot interval, the minimum of 5 values is often enough. These are typically equally spaced between adjacent knots. The second column contains the weights. For Simpson's rule, these are proportional to 1, 4, 2, 4, ..., 2, 4, 1.

values   A list, with entries containing the values of the basis function derivatives starting with 0 and going up to the highest derivative needed. The values correspond to quadrature points in quadvals. It is up to the user to decide whether or not to multiply the derivative values by the square roots of the quadrature weights so as to make numerical integration a simple matrix multiplication. Values are checked against quadvals to ensure the correct number of rows, and against nbasis to ensure the correct number of columns; values contains values of basis functions and derivatives at quadrature points weighted by square root of quadrature weights. These values are only generated as required, and only if the quadvals is not matrix("numeric",0,0).

basisvalues   A list of lists. This is designed to avoid evaluation of a basis system repeatedly at a set of argument values. Each sublist corresponds to a specific set of argument values, and must have at least two components, which may be named as you wish. The first component in an element of the list vector contains the argument values. The second component is a matrix of values of the basis functions evaluated at the arguments in the first component. Subsequent components, if present, are matrices of values of their derivatives up to a maximum derivative order. Whenever function getbasismatrix is called, it checks the first list in each row to see first if the number of argument values corresponds to the size of the first dimension, and if this test succeeds, checks that all of the argument values match.

The names of the suffixes in R or the functions in Matlab that either extract or insert component information into a basis object are shown in Table 3.1.

Table 3.1 The methods for extracting and modifying information in a `basisfd` (R) or `basis` (Matlab) object

R suffix	Matlab function	
`$nbasis`	`getnbasis`	`putnbasis`
`$dropind`	`getdropind`	`putdropind`
`$quadvals`	`getquadvals`	`putquadvals`
`$basisvalues`	`getbasisvalues`	`putbasisvalues`
`$values`	`getvalues`	`putvalues`

3.7 Some Things to Try

1. Work the examples in the help page for `create.fourier.basis` and `create.bspline.basis`. What do these examples tell you about these alternative basis systems?
2. Generate a B-spline basis. Follow these steps:

   a. Decide on the range, such as perhaps [0,1].
   b. Choose an order, such as four.
   c. Specify the number of basis functions. The more you specify, the more variability you can achieve in the function. As a first choice, 23 might be reasonable; for order four splines, this places by default knots at $0, 0.05, 0.10, \ldots,$ $0.90, 0.95$ and 1 over [0,1].
   d. Plot the basis to see how it looks using the `plot` command.
   e. Now evaluate and plot a few derivatives of the basis functions to see how their smoothness diminishes with each successive order of derivative.

Chapter 4
How to Build Functional Data Objects

We saw in the last chapter that functions are built up from basis systems $\phi_1(t), \ldots,$ $\phi_K(t)$ by defining the linear combination

$$x(t) = \sum_{k=1}^{K} c_k \phi_k(t) = \mathbf{c}' \phi(t).$$

That chapter described how to build a basis system. Now we take the next step, defining a functional data object by combining a set of coefficients c_k (and other useful information) with a previously specified basis system.

4.1 Adding Coefficients to Bases to Define Functions

4.1.1 Coefficient Vectors, Matrices and Arrays

Once we have selected a basis, we have only to supply coefficients in order to define an object of the *functional data* class (with class name fd).

If there are K basis functions, we need a coefficient vector of length K for each function that we wish to define. If only a single function is defined, then the coefficients are loaded into a vector of length K or a matrix with K rows and one column. If N functions are needed, say for a sample of functional observations of size N, we arrange these coefficient vectors in a K by N matrix. If the functions themselves are multivariate of dimension m, as would be the case, for example, for positions in three-dimensional space ($m = 3$), then we arrange the coefficients into a three-dimensional array of dimensions K, N, and m, respectively. (A single multivariate function is defined with a coefficient array with dimensions $K, 1$, and m; see Section 2.2 for further information on this case.) That is, the dimensions are in the order "number of basis functions," "number of functions or functional observations" and "number of dimensions of the functions."

J.O. Ramsay et al., *Functional Data Analysis with R and MATLAB*, Use R,
DOI: 10.1007/978-0-387-98185-7_4,
© Springer Science + Business Media, LLC 2009

Here is the command that creates a functional data object using the basis with name daybasis65 that we created in the previous chapter, with the coefficients for mean temperature for each of the 35 weather stations organized into the 65 by 35 matrix coefmat:

```
tempfd = fd(coefmat, daybasis65)
```

You will seldom need to use the fd function explicitly because other functions call it after computing coefmat as a representation of functional data in terms of the specified basis set. We will discuss some of these functions briefly later in this chapter and in more detail in the next.

4.1.2 Labels for Functional Data Objects

Let us take a moment here to reflect on what functional data objects mean. Functional data objects represent functions, and functions are one-to-one mappings or relationships between values in a *domain* and values in a *range*. In the language of graphics, the domain values are points on the horizontal coordinate or *abscissa*, and the range values are points in a vertical coordinate or *ordinate*. For the purpose of this book, we consider mostly one-dimensional domains, such as time, but we do allow for the possibility that the range space of multidimensional, such as (X,Y,Z) triples for the coordinates of points in a three-dimensional space. Finally, we also allow for the possibility of multiple or replicated functions.

Adding labels to functional data objects is a convenient way to supply the information needed for graphical displays. Specialized plotting functions that the code supplies in either language can look for these labels, and if they are present, place them where appropriate for various kinds of plots. The component for labels for functional data objects is called fdnames.

If we want to supply labels, we will typically need three, and they are, in order:

1. A label for the domain, such as 'Time', 'Day', and so on.
2. A label for the replication dimension, such as as 'Weather station', 'Child', etc.
3. A label for the range, such as "Temperature (deg. C)', 'Space', etc.

We refer to these three labels as the *generic labels* for the functional data object.

In R, we supply labels in a list object of length three. An empty version of such a list can be set up by the command

```
fdnames = vector("list", 3)
```

The corresponding object in Matlab is a cell array of length three, which may be set up by

```
fdnames = cell(1,3)
```

In addition to generic labels for each dimension of the data, we may also want, for the range and/or for the replication dimension, to supply sets of labels, each label applying to a specific dimension or replicate. For example, for the gait data, we may want a label such as "Angle" to be common or generic to the two observed angles, but in addition require two labels such as "Knee" and "Hip" to distinguish which angle is being plotted. Similarly, in addition to "Weather Station" to describe generically the replication dimension for the weather data as a whole, we probably want to supply names for each weather station. Thus, labels for replicates and variables have the potential to have two levels, a generic level and a specific level. Of course, if there is only one dimension for range or only one replicate, a two-level labels structure of this nature would usually be superfluous.

In the simple case where a dimension only needs a single name, labels are supplied as strings having the class `character` in R or `char` in Matlab. For example, we may supply only a common name such as "Child" for the replication dimension of the growth data, and "Height(cm)" for the range, combined with "Age (years)" for the domain. Here is a command that sets up these labels in R directly, without bothering to set up an empty list first,

```
fdnames = list("Age (years)", "Child", "Height (cm)")
```

or, assuming that the empty list has already been defined:

```
fdnames[[1]] = "Age (years"
fdnames[[2]] = "Child"
fdnames[[3]] = "Height (cm)"
```

Since Matlab accesses cell array elements by curly brackets the expressions are

```
fdnames{1} = 'Age (years)'
fdnames{2} = 'Child'
fdnames{3} = 'Height (cm)'
```

However, when the required label structure for either the replication or the range dimension is two-level, we take advantage of the fact that the elements of a list in R can be character vectors or lists, and entries in cell arrays in Matlab can be cell arrays. We deal with the two languages separately in the following two paragraphs.

In R, generic and specific names can be supplied by a named list. The common or generic label is supplied by the name of the list and the individual labels by the entry of the list, this entry being of either the character or list class. Take weather stations for the weather data, for example. The second element is itself a list, defined perhaps by the commands

```
station = vector("list", 35)
station[[ 1]] = "St. Johns"
            .

            .

            .
station[[35]] = "Resolute"
```

A command to set up a labels list for the daily temperature data might be

```
fdnames = list("Day",
               "Weather Station" = station,
               "Mean temperature (deg C)")
```

Notice that the `names` attribute of a list entry can be a quoted string containing blanks, such as what we have used here. The other two names, `argname` and `varname`, will only be used if the entry is `NULL` or `""` or, in the case of variable name, if the third list entry contains a vector of names of the same length as the number of variables. The code also checks that the number of labels in the label vector for replications equals the number of replications and uses the `names` value if this condition fails.

Matlab does not have an analogue of the `names` attribute in R, but each entry in the cell array of length three can itself be a cell array. If the entry is either a string or a cell array whose length does not match the required number of labels, then the Matlab plotting functions will find in this entry a generic name common to all replicates or variables. But if the entry for either the replicates or variables dimension is a cell array of length two, then the code expects the generic label in the first entry and a character matrix of the appropriate number of rows in the second. The weather station example above in Matlab becomes

```
station=cell(1,2);
station{1} = 'Weather Station';
station{2} = ['St. Johns      ';
              'Charlottetown';
                      .
                      .
                      .
              'Resolute       '];
```

Note that a series of names are stored as a matrix of characters, so that enough trailing blanks in each name must be added to allow for the longest name to be used.

4.2 Methods for Functional Data Objects

As for the basis class, there are similar generic functions for printing, summarizing and testing for class and identity for functional data objects.

There are, in addition, some useful methods for doing arithmetic on functional data objects and carrying out various transformations. For example, we can take the sum, difference, power or pointwise product of two functions with commands like

```
fdsumobj = fdobj1 + fdobj2
fddifobj = fdobj1 - fdobj2
fdprdobj = fdobj1 * fdobj2
fdsqrobj = fdobj^2
```

One can, as well, substitute a scalar constant for either argument in the three arithmetic commands. We judged pointwise division to be too risky since it is difficult to detect if the denominator function is nonzero everywhere. Similarly,

```
fdobj^a
```

may produce an error or nonsense if a is negative and fdobj is possibly zero at some point.

Beyond this, the results of multiplication and exponentiation may not be what one might naively expect. For example, the following produces a straight line from (-1) to 2 with a linear spline basis:

```
tstFn0 <- fd(c(-1, 2), create.bspline.basis(norder=2))
plot(tstFn0)
```

However,

```
tstFn0^2
```

is not a parabola but a straight line that approximates this parabola over rangeval using the same linear basis set. We get a similar approximation from tstFn0*tstFn0, but it differs in the third significant digit.

What do we get from

```
tstFn0^(-1)?
```

The result may be substantially different from what many people expect. These are known "infelicities" in fda, which the wise user will avoid. Using cubic or higher-order splines with basis sets larger than in this example will reduce substantially these problems in many but not all cases.

The mean of a set of functions is achieved by a command like

```
fdmeanobj = mean(fdobj)
```

Similarly, functions are summed by the sum function. As the software evolves, we expect that other useful methods will be added (and infelicities further mitigated).

We often want to work with the values of a function at specified values of argument t, stored, say, in vector tvec. The evaluation function comparable to that used in Chapter 3 for basis functions is eval.fd in R and eval_fd in Matlab. For example, we could evaluate functional data object thawfd at times in vector day.5 by the R command

```
thatvec = eval.fd(tvec, thawfd)
```

The same command can be used to evaluate a derivative of thawfd by supplying the index of the derivative as the third argument. The second derivative of thawfd is evaluated by

```
D2thatvec = eval.fd(tvec, thawfd, 2)
```

More generally, if Lfdobj is an object of the linear differential operator Lfd class, defined in Section 4.4, then

```
Lthatvec = eval.fd(tvec, thawfd, Lfdobj)
```

evaluates the result of applying this operator to `thawfd` at the argument values in `tvec`. (In R, `predict.fd` provides a wrapper for `eval.fd` with a different syntax that is more consistent with the standard R generic `predict` function.)

Plotting functions can be as simple as using the command `plot(tempfd)`. Again, Matlab and R use the class name to find the plotting function that is appropriate to what is being plotted, which in this case is an object of the `fd` class. The functional data version of the `plot` function can also use most of the optional arguments available in the standard plotting function for controlling line color, style and width; axis limits; and so forth.

Here is a set of R commands that plot the mean temperature curves for the Canadian weather data after loading the `fda` package. First, we set up the midday times for each of the days in years that are not leap years.

```
daytime = (1:365)-0.5
```

In this book we will find more of interest in the winter months, so we highlight these by rearranging the standard year to run from July 1 to June 30.

```
JJindex = c(182:365, 1:181)
tempmat = daily$tempav[JJindex,]
```

Next we set up a Fourier basis with 65 basis functions, as we did in Chapter 3.

```
tempbasis = create.fourier.basis(c(0,365),65)
```

Now we use the main smoothing function that we will study in Chapter 5 to set up the functional data object `tempfd`, and install names for the three dimensions of the object.

```
tempfd = smooth.basis(daytime, tempmat, tempbasis)$fd
tempfd$fdnames = list("Day (July 2 to June 30)",
                      "Weather Station",
                      "Mean temperature (deg. C)")
```

Finally we plot the 35 mean temperature functions, shown in Figure 4.1, using the optional standard plotting arguments `col` and `lty` to control the color and line style, respectively.

```
plot(tempfd, col=1, lty=1)
```

Lines defined by functional data objects can be added to an existing plot by the `lines` function in R or the `line` function in Matlab.

4.2.1 Illustration: Sinusoidal Coefficients

We pointed out in Chapter 3 that curves defined by B-spline bases tend to follow the same track as their coefficients. Here is an example. This R code sets up a coefficient

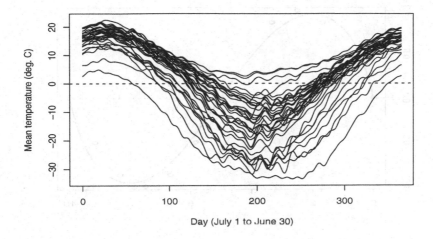

Fig. 4.1 Mean temperature curves estimated by R command `tempfd = smooth.basis(daytime, tempmat, tempbasis)$fd`, and plotted by command `plot(tempfd)`.

vector of length 13 consisting of values of a sinc wave at equally spaced values over its cycle, and then uses these along with the basis system plotted in Figure 3.1 to define a functional data object. Both the defined curve and the sine-valued coefficients are plotted in Figure 4.2. The curve is not a perfect rendition of a spline, but it is surprisingly close.

```
basis13 = create.bspline.basis(c(0,10), 13)
tvec = seq(0,1,len=13)
sinecoef = sin(2*pi*tvec)
sinefd = fd(sinecoef, basis13, list("t","","f(t)"))
op = par(cex=1.2)
plot(sinefd, lwd=2)
points(tvec*10, sinecoef, lwd=2)
par(op)
```

4.3 Smoothing Using Regression Analysis

The topic of smoothing data will be taken up in detail in Chapter 5. However, we can sometimes get good results without more advanced smoothing machinery simply by keeping the number of basis functions small relative to the amount of data being approximated.

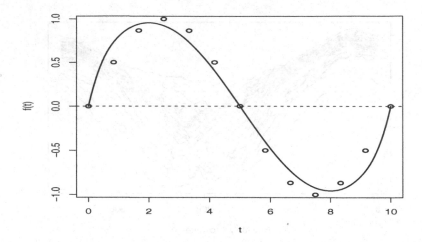

Fig. 4.2 The 13 spline basis functions defined in Figure 3.1 are combined with coefficients whose values are sinusoidal to construct the functional data object plotted as a solid line. The coefficients themselves are plotted as circles.

4.3.1 Plotting the January Thaw

Canadians love to talk about the weather, and especially in midwinter when the weather puts a chill on many other activities. The January thaw is eagerly awaited, and in fact the majority of Canadian weather stations show clear evidence of these few days of relief. The following code loads 34 years of daily temperature data for Montreal, extracts temperatures for January 16th to February 15th and plots their mean, shown in Figure 4.3.

```
# This assumes the data are in "MtlDaily.txt"
# in the working directory getwd()
MtlDaily = matrix(scan("MtlDaily.txt",0),34,365)
thawdata = t(MtlDaily[,16:47])
daytime  = ((16:47)+0.5)
par(cex=1.2)
plot(daytime, apply(thawdata,1,mean), "b", lwd=2,
     xlab="Day", ylab="Temperature (deg C)")
```

We can fit these data by regression analysis by using a matrix of values of a basis system taken at the times in vector daytime. Here we construct a basis system over the interval [16,48] using seven cubic B-splines, and evaluate this basis at these points to produce a 32 by 7 matrix. By default the knots are equally spaced over this interval.

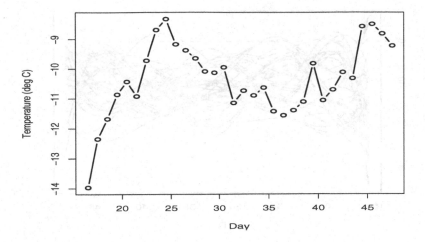

Fig. 4.3 Temperatures at Montreal from January 16 to February 15 averaged over 1961 to 1994.

```
thawbasis     = create.bspline.basis(c(16,48),7)
thawbasismat = eval.basis(daytime, thawbasis)
```

Now we can compute coefficients for our functional data object by the usual equations for regression coefficients, $\mathbf{b} = (\mathbf{X'X})^{-1}\mathbf{X'y}$, and construct a functional data object by combining them with our basis object. A plot of these curves is shown in Figure 4.4 and, sure enough, we do see a fair number of them peaking between January 20 and 25, and a few others with later peaks as well.

```
thawcoef = solve(crossprod(thawbasismat),
                 crossprod(thawbasismat,thawdata))
thawfd = fd(thawcoef, thawbasis,
        list("Day", "Year", "Temperature (deg C)"))
plot(thawfd, lty=1, lwd=2, col=1)
```

We can use these objects to illustrate two useful tools for working with functional data objects. We often want to compare a curve to the data from which it was estimated. In the following command we use function plotfit.fd to plot the data for 1961 along with corresponding B-spline fit. The command also illustrates the possibility of using subscripts on functional data objects. The result is shown in Figure 4.5, where the fit suggests a thaw before January 15 and another in early February. The legend on the plot indicates that the standard deviation of the variation of the actual temperatures around the curve is four degrees Celsius.

```
plotfit.fd(thawdata[,1], daytime, thawfd[1],
           lty=1, lwd=2)
```

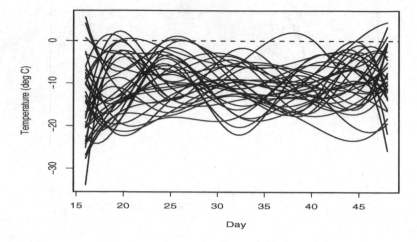

Fig. 4.4 Functional versions of temperature curves for Montreal between January 16 and February 15. Each curve corresponds to one of the years from 1960 to 1994.

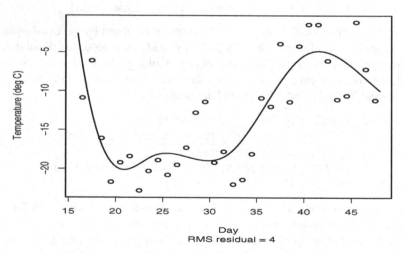

Fig. 4.5 The temperature curve for 1961 along with the actual temperatures from which it is estimated.

4.4 The Linear Differential Operator or **Lfd** Class

We noted in Chapter 1 that the possibility of using a derivative of a function is perhaps the most distinctive feature of functional data analysis. For example, we will take advantage of the information in derivatives in Chapter 5 to customize our definition of what we mean by a "smooth" function. Our discussion in Section 1.4 also implied that the concept of a "derivative" could itself be extended by proposing linear combinations of derivatives, called *linear differential operators*.

Smoothing is supported using the Lfd class that expresses the concept of a linear differential operator. An important special case is the *harmonic acceleration* operator that we will use extensively with Fourier basis functions to smooth periodic data.

The notation Lx refers to the application of a linear differential operator L to a function x. This might be something as basic as acceleration, $Lx = D^2x$, as moderately sophisticated as harmonic acceleration $L = \omega^2 D + D^3$, or as general as

$$Lx(t) = \beta_0(t)x(t) + \beta_1(t)Dx(t) + \ldots + \beta_{m-1}(t)D^{m-1}x(t) + D^m x(t) \qquad (4.1)$$

where the known *linear differential operator coefficient functions* $\beta_j(t), j = 0, \ldots,$ $m - 1$ are either constants or functions.

How do we express this idea in code so as to permit the use of the full potential residing in (4.1)? How do we even do this for harmonic acceleration?

The Lfd class is defined by a constructor function Lfd that takes as its input two arguments:

nderiv   The highest order m of the derivative in (4.1).

bwtlist   A list object in R or a one-dimensional cell array object in Matlab of length m. This object contains the coefficient functions β_j defining the operator. If a coefficient function varies over t, these will be functional data objects with a single replication. But if the coefficient is constant, including zero, the corresponding entry will be that constant.

For example, consider the harmonic acceleration object. Here $m = 3$, and $\beta_0 = \beta_2 = 0$ while $\beta_1 = \omega^2$. In R we could define the harmonic acceleration Lfd object harmaccelLfd in this way:

```
betalist = vector("list", 3)
betalist[[1]] = fd(0, thawconst.basis)
betalist[[2]] = fd(omega^2, thawconst.basis)
betalist[[3]] = fd(0, thawconst.basis)
harmaccelLfd = Lfd(3, betalist)
```

This is a bit cumbersome for the majority of situations where the differential operator is just a power of D or where all the coefficients β_j are constants. Consequently, we have two functions, int2Lfd and vec2Lfd, to deal with these simpler situations:

```
accelLfd      = int2Lfd(2)
```

```
harmaccelLfd = vec2Lfd(c(0,omega^2,0), c(0, 365))
```

The commands `class(accelLfd)` and `class(harmaccelLfd)` will produce `Lfd` as output.

We are now equipped to evaluate the result of applying any linear differential operator to a functional data object. We can illustrate this by applying an appropriately defined harmonic acceleration operator to temperature curves in functional data object `tempfd`:

```
Ltempmat = eval.fd(daytime, tempfd, harmaccelLfd)
```

This was used to prepare Figure 1.12.

A functional data object for the application of a linear differential operator to an existing functional data object is created by the `deriv.fd` function in R or the `deriv_fd` function in Matlab. The first argument is the functional data object for which the derivative is required and the second is either a nonnegative integer or a linear differential operator object. For example,

```
D2tempfd = deriv.fd(tempfd, 2)
Ltempfd  = deriv.fd(tempfd, harmaccelLfd)
```

4.5 Bivariate Functional Data Objects: Functions of Two Arguments

The availability of a sample of N curves makes us wonder how they vary among themselves. The analogue of the correlation and covariance matrices in the multivariate context are the correlation and covariance functions or surfaces, $\rho(s,t)$ and $\sigma(s,t)$. The value $\rho(s,t)$ specifies the correlation between the values $x(s)$ and $x(t)$ over a sample or population of curves, and similarly for $\sigma(s,t)$. This means that we also need to be able to define functions of two arguments, in this case s and t. We will need this capacity elsewhere. Certain types of functional regression require bivariate regression coefficient functions.

The bivariate functional data class with name `bifd` is designed to do this. Objects of this class are created in much the same way as `fd` objects, but this now requires two basis systems and a matrix of coefficients for a single such object. In mathematical notation, we define an estimate of a bivariate correlation surface as

$$r(s,t) = \sum_{1}^{K}\sum_{1}^{L} b_{k,\ell}\phi_k(s)\psi_\ell(t) = \phi'(s)\mathbf{B}\psi(t), \qquad (4.2)$$

where $\phi_k(s)$ is a basis function for variation over s and $\psi_\ell(t)$ is a basis function for variation over t. The following command sets up such a bivariate functional object:

```
corrfd = bifd(corrmat, sbasis, tbasis)
```

However, situations where you would have to set up bivariate functional data objects are rare, since most of these are set up by R or Matlab functions, `var.fd` and `var_fd` in R and Matlab, respectively. We will use these functions in Chapter 6.

4.6 The Structure of the `fd` and `Lfd` Classes

To summarize the most important points of this chapter, we give here the arguments of the constructor function `fd` for an object of the `fd` class.

coef   A vector, matrix, or three-dimensional array of coefficients. The first dimension (or elements of a vector) corresponds to basis functions. A second dimension corresponds to the number of functional observations, curves or replicates. If `coef` is a three-dimensional array, the third dimension corresponds to variables for multivariate functional data objects.

basisobj   A functional basis object defining the basis.

fdnames   A list of length three, each member potentially being a string vector containing labels for the levels of the corresponding dimension of the data. The first dimension is for argument values and is given the default name `"time"`. The second is for replications and is given the default name `"reps"`. The third is for functions and is given the default name `"values"`.

The arguments of the constructor function Lfd for objects of the linear differential operator class are

nderiv   A nonnegative integer specifying the order m of the highest order derivative in the operator.

bwtlist   A list of length m. Each member contains a functional data object that acts as a weight function for a derivative. The first member weights the function, the second the first derivative, and so on up to order $m - 1$.

4.7 Some Things to Try

1. Generate a random function using a B-spline basis. Follow these steps:

   a. Decide on the range, such as [0,1].
   b. Choose an order, such as four for cubic splines.
   c. Specify the number of basis functions. The more you specify, the more variability you can achieve in the function. As a first choice, 23 might be reasonable; for order four splines over [0, 1], this places by default knots at $0, 0.05, 0.10, \ldots, 0.90, 0.95$ and 1.

d. Now set up the basis function system in the language you are working with. Plot the basis to see how it looks using the `plot` command (as described in the previous chapter on basis sets).

e. Next define a vector of random coefficients using your language's normal random number generator. These can vary about zero as a mean, but you can also vary them around some function, such as $\sin(2\pi t)$ over [0,1]. If you use a trend, because of the unit sum property of B-splines described above, the function you define will also vary around this trend. You may want to play around their standard deviation as a part of this exercise.

f. Finally, set up a functional data object having a single function using the `fd` command.

2. Plot this function using the `plot` command.

3. Plot both the function and the coefficients on the same graph. To plot the coefficients for order four splines, plot all but the second and third in from each end against knot locations. For example, if you have 23 basis functions, and hence 23 coefficients, plot coefficients 1, 4, 5, and so on up to 20, and then the 23rd. The 21 knots (including end points) are equally spaced by default. At the same time, evaluate the function using the `eval.fd` (R) or `eval_fd` (Matlab) function at a fine mesh of values, such as 51 equally spaced values. Plot these values over the coefficients that you have just plotted. Compare the trend in the coefficients and the curve. If you specified a mean function for the random coefficients, you might want to add this to the plot as well.

4. You might want to extend this exercise to generating N random functions, and plot all of them simultaneously to see how much variation there is from curve to curve. This will, of course, depend on the standard deviation of the random coefficients that you use.

5. Why not also plot the first and second derivatives of these curves, evaluated again using the `eval.fd` function and specifying the order of derivative as the third argument. You might want to compare the first derivative with the difference values for the coefficients.

Chapter 5
Smoothing: Computing Curves from Noisy Data

The previous two chapters have introduced the Matlab and R code needed to specify basis function systems and then to define curves by combining these coefficient arrays. For example, we saw how to construct a basis object such as `heightbasis` to define growth curves and how to combine it with a matrix of coefficients such as `heightcoef` so as to define growth functional data objects such as were plotted in Figure 1.1.

We now turn to methods for computing these coefficients with more careful consideration of measurement error. For example, how do we compute these coefficients to obtain an optimal fit to data such as the height measurements for 54 girls in the Berkeley growth study stored in the 31 by 54 matrix that we name `heightmat`? Or how do we replace the rather noisy mean daily precipitation observations by smooth curves?

Two strategies are discussed. The simplest revisits the use of regression analysis that concluded Chapter 4, but now uses a special function for this purpose. The second and more elaborate strategy aims to miss nothing of importance in the data by using a powerful basis expansion, but avoids overfitting the data by imposing a penalty on the "roughness" of the function, where the meaning of "rough" can be adapted to special features of the application from which the data were obtained.

5.1 Regression Splines: Smoothing by Regression Analysis

We tend, perhaps rather too often, to default to defining data fitting as the minimization of the sum of squared errors or residuals,

$$\text{SSE}(x) = \sum_{j}^{n} [y_j - x(t_j)]^2. \tag{5.1}$$

When smoothing function x is defined as a basis function expansion (3.1), the least-squares estimation problem becomes

J.O. Ramsay et al., *Functional Data Analysis with R and MATLAB*, Use R,
DOI: 10.1007/978-0-387-98185-7_5,
© Springer Science + Business Media, LLC 2009

$$\text{SSE}(\mathbf{c}) = \sum_{j}^{n}[y_j - \sum_{k}^{K} c_k \phi_k(t_j)]^2 = \sum_{j}^{n}[y_j - \phi(t_j)'\mathbf{c}]^2. \qquad (5.2)$$

The approach is motivated by the error model

$$y_j = x(t_j) + \varepsilon_j = \mathbf{c}'\phi(t) + \varepsilon_j = \phi'(t_j)\mathbf{c} + \varepsilon_j \qquad (5.3)$$

where the true errors or residuals ε_j are statistically independent and have a normal or Gaussian distribution with mean 0 and constant variance. Of course, if we look closely, we often see that this error model is too simple. Nevertheless, the least-squares estimation process can be defended on the grounds that it tends to give nearly optimal answers relative to "best" estimation methods so long as the true error distribution is fairly short-tailed and departures from the other assumptions are reasonably mild.

Readers will no doubt recognize (5.3) as the standard regression analysis model, along with its associated least-squares solution. Using matrix notation, let the n-vector $\mathbf{y}$ contain the n values to be fit, vector ε contain the corresponding true residual values, and n by k matrix Φ contain the basis function values $\phi_k(t_j)$. Then we have

$$\mathbf{y} = \Phi\mathbf{c} + \varepsilon$$

and the least-squares estimate of the coefficient vector $\mathbf{c}$ is

$$\hat{\mathbf{c}} = (\Phi'\Phi)^{-1}\Phi'\mathbf{y}. \qquad (5.4)$$

R and Matlab already have the capacity to smooth data through their functions for regression analysis. Here is how we can combine these functions with the basis creation functions available in the `fda` package. Suppose that we want a basis system for the growth data with $K = 12$ basis functions using equally spaced knots. This can be accomplished in R with the following command:

```
heightbasis12 = create.bspline.basis(c(1,18), 12, 6)
```

If we evaluate the basis functions at the ages of measurement in vector object `age` by the command `basismat = eval.basis(age, heightbasis12)` (in R), then we have a 31 by 12 matrix of covariate or design values that we can use in a least-squares regression analysis defined by commands such as

```
heightcoef = lsfit(basismat, heightmat,
                   intercept=FALSE)$coef
heightcoef = basismat\heightmat
```

in R and Matlab, respectively. Spline curves fit by regression analysis are often referred to as *regression splines* in statistical literature.

However, the function `smooth.basis` (R) and `smooth_basis` (Matlab) are provided to produce the same results as well as much more without the need to explicitly evaluate the basis functions, through the R command

```
heightList = smooth.basis(age, heightmat,
```

```
                              heightbasis12)
```

and the Matlab version

```
[fdobj, df, gcv, coef, SSE, penmat, y2cMap] = ...
    smooth_basis(age, heightmat, heightbasis12);
```

The R function smooth.basis returns an object heightlist of the list class, and the Matlab function smooth_basis returns all seven of its objects as an explicit sequence of variable names surrounded by square brackets. However, if we just wanted the first three returned objects as separate objects, in R we would have to extract them individually:

```
heightfd   = heightList$fd
height.df  = heightList$df
height.gcv = heightList$gcv
```

In Matlab, we would just request only the first three objects:

```
[fdobj, df, gcv] = ...
    smooth_basis(age, heightmat, heightbasis12);
```

In any case, the three most important returned objects are the following, where the names in bold type are used in each language to retrieve the objects:

fd   An object of class fd containing the curves that fit the data.

df   The degrees of freedom used to define the fitted curves.

gcv   The value of the generalized cross-validation criterion: a measure of lack of fit discounted for degrees of freedom. If there are multiple curves, a vector is returned containing gcv values for each curve. (See Ramsay and Silverman (2005) for details.)

Notice that the coefficient estimate $\hat{c}$ in (5.4) is obtained from the data in the vector **y** by multiplying this vector by a matrix, to which we give the text name y2cMap. We will use this matrix in many places in this book where we need to estimate the *variability* in quantities determined by $\hat{c}$, so we here give it a name:

$$\text{y2cMap} = (\Phi'\Phi)^{-1}\Phi' \text{ so that } \hat{c} = \text{y2cMap } \mathbf{y}. \qquad (5.5)$$

Here is the corresponding R code for computing this matrix for the growth data:

```
age = growth$age
heightbasismat = eval.basis(age, heightbasis12)
y2cMap = solve(crossprod(heightbasismat),
                     t(heightbasismat))
```

In Matlab this last command would be

```
y2cMap = (heightbasismat'*heightbasismat) \ ...
        heightbasismat';
```

This code for the mapping matrix y2cMap only applies to regression-based smoothing. More general expressions for y2cMap include other term(s) that disappear with zero smoothing. This is important because as we change the smoothing, y2cMap changes, but $\hat{c}$ is still the product of y2cMap, however changed, and the data.

While we are at it, we also will need what is often called the "hat" matrix, denoted by **H**. This maps the data vector into the vector of fitted values

$$\mathbf{H} = \Phi(\Phi'\Phi)^{-1}\Phi' \text{ so that } \hat{\mathbf{y}} = \mathbf{Hy}. \tag{5.6}$$

The regression approach to smoothing data only works if the number K of basis functions is substantially smaller than the number n of sampling points. With the growth data, it seems that roughly $K = 12$ spline basis functions are needed to adequately smooth the growth data. Larger values of K will tend to undersmooth or overfit the data. Interestingly, after over a century of development of parametric growth curve models, the best of these also use about 12 parameters in this example.

Although regression splines are often adequate for simple jobs where only curve values are to be used, the instability of regression spline derivative estimates at the boundaries is especially acute. The next section describes a more sophisticated approach that can produce much better derivative results and also allows finer control over the amount of smoothing.

5.2 Data Smoothing with Roughness Penalties

The roughness penalty approach uses a large number of basis functions, possibly extending to one basis function per observation and even beyond, but at the same time imposing smoothness by penalizing some measure of function complexity. For example, we have already in the last chapter defined a basis system for the growth data called heightbasis that has 35 basis functions, even though we have only 31 observations per child. Would using such a basis system result in overfitting the data, as well as singularity problems on the computational side? That answer is, "Not if a positive penalty is applied to the degree to which the fit is not smooth."

5.2.1 Choosing a Roughness Penalty

We define a measure of the *roughness* of the fitted curve, and then minimize a fitting criterion that trades off curve roughness against lack of data fit.

Here is a popular way to quantify the notion "roughness" of a function. The square of the second derivative $[D^2x(t)]^2$ of a function x at argument value t is often called its *curvature* at t, since a straight line, which we all tend to agree has no curvature, has second derivative zero. Consequently, a measure of a function's roughness is the *integrated squared second derivative* or *total curvature*

$$\text{PEN}_2(x) = \int [D^2 x(t)]^2 \, dt . \tag{5.7}$$

(Unless otherwise stated, all integrals in this book are definite integrals over the range of t.)

Penalty terms such as $\text{PEN}_2(x)$ provide smoothing because wherever the function is highly variable, the square of the second derivative $[D^2 x(t)]^2$ is large. We can apply this concept to derivative estimation as well. If we are interested in the second derivative $D^2 x$ of x, chances are that we want it to appear to be smooth. This suggests that we ought to penalize the curvature of the second derivative, that is, use the roughness measure

$$\text{PEN}_4(x) = \int [D^4 x(t)]^2 \, dt . \tag{5.8}$$

But is "roughness" always related to the second derivative? Thinking a bit more broadly, we can define roughness as the extent to which a function departs from some baseline "smooth" behavior. For periodic functions of known period that can vary in level, such as mean temperature curves, the baseline behavior can be considered to be shifted sinusoidal variation,

$$x(t) = c_0 + a_1 \sin \omega t + b_1 \cos \omega t, \tag{5.9}$$

that is, represented by the first three terms in the Fourier series for some known $\omega = 2\pi/T$. If we compute $\omega^2 Dx + D^3 x$ for such a simple function, we find that the result is exactly 0. We refer to the *differential operator* $L = \omega^2 D + D^3$ in Ramsay and Silverman (2005) as the *harmonic acceleration operator*. What happens when we apply this *harmonic acceleration operator* to higher-order terms in a Fourier series:

$$L[a_j \sin j\omega t + b_j \cos j\omega t] = \omega^2 j(1 - j^2)[a_j \cos j\omega t - b_j \sin j\omega t]. \tag{5.10}$$

This expression is 0 for $j = 1$ and increases with the cube of j. This property suggests that the integral of the square of this *harmonic acceleration operator* may be a suitable measure of roughness for periodic data like the temperatures curves:

$$\text{PEN}_L(x) = \int [Lx(s)]^2 \, ds . \tag{5.11}$$

When used on a finite Fourier series, this expression is proportional to $[j^2(1 - j^2)^2]$. Thus, the term with $j = 1$ does not get penalized at all, and higher-order terms in the Fourier approximation receive substantially higher penalties.

Whatever roughness penalty we use, we add some multiple of it to the error sum of squares to define the compound fitting criterion. For example, using $\text{PEN}_2(x)$ gives us the following:

$$F(\mathbf{c}) = \sum_j [y_j - x(t_j)]^2 + \lambda \int [D^2 x(t)]^2 dt, \tag{5.12}$$

where $x(t) = \mathbf{c}' \boldsymbol{\phi}(t)$. The *smoothing parameter* λ specifies the emphasis on the second term penalizing curvature relative to goodness of fit quantified in the sum of squared residuals in the first term. As λ moves from 0 upward, curvature becomes increasingly penalized. With λ sufficiently large, $D^2 x$ will be essentially 0. This in turn implies that x will be essentially a straight line = polynomial of degree one, order two, except possibly at a finite number of isolated points such as join points or knots of a B-spline. At the other extreme, $\lambda \to 0$ leaves the function x free to fit the data as closely as possible with the selected basis set, sometimes at the expense of some fairly wild variations in the approximating function.

It is usually convenient to plot and modify λ on a logarithmic scale. More generally, the use of a differential operator L to define roughness will result in $\lambda \to \infty$ forcing the fit to approach more and more closely a solution to the differential equation $Lx = 0$. If $L = D^m$, this solution will be a polynomial of order m (i.e., degree $m - 1$). For the harmonic acceleration operator, this solution will be of the form (5.9). In this way, we can achieve an important new form of control over the smoothing process, namely by having the capacity to define the concept "smooth" in a way that is appropriate to the application.

5.2.2 The Roughness Penalty Matrix R

We can now provide an explicit form of the estimate of the coefficient vector $\hat{\mathbf{c}}$ for roughness penalty smoothing that is the counterpart of (5.4) for regression smoothing. The general version of the roughness penalized fitting criterion (5.12) is

$$F(\mathbf{c}) = \sum_j [y_j - x(t_j)]^2 + \lambda \int [Lx(t)]^2 dt. \tag{5.13}$$

If we substitute the basis expansion $x(t) = \mathbf{c}' \boldsymbol{\phi}(t) = \boldsymbol{\phi}'(t)\mathbf{c}$ into this equation, we get

$$F(\mathbf{c}) = \sum_j [y_j - \boldsymbol{\phi}'(t_j)\mathbf{c}]^2 + \lambda \mathbf{c}' [\int L\boldsymbol{\phi}(t) L\boldsymbol{\phi}'(t) dt] \mathbf{c}. \tag{5.14}$$

Now we define the order K symmetric roughness penalty matrix as

$$\mathbf{R} = \int \boldsymbol{\phi}(t) \boldsymbol{\phi}'(t) dt. \tag{5.15}$$

With this defined, it is a relatively easy exercise in matrix algebra to work out that

$$\hat{\mathbf{c}} = (\boldsymbol{\Phi}'\boldsymbol{\Phi} + \lambda \mathbf{R})^{-1} \boldsymbol{\Phi}' \mathbf{y}. \tag{5.16}$$

From here we can define the matrix y2cMap that we will use in Chapter 6 for computing confidence regions about estimated curves:

$$y2cMap = (\Phi'\Phi + \lambda R)^{-1}\Phi'. \tag{5.17}$$

The corresponding hat-matrix is now

$$\mathbf{H} = \Phi(\Phi'\Phi + \lambda R)^{-1}\Phi'. \tag{5.18}$$

But how is one to compute matrix $\mathbf{R}$ in either language? This is taken care of in the function eval.penalty in R and eval_penalty in Matlab. These functions require two arguments:

basisobj   A functional basis object of the basisfd class in R and basis class in Matlab.
Lfdobj   A linear differential operator object of the Lfd class.

In the case of the harmonic accelerator operator, we can calculate the roughness penalty matrix Rmat in R by

```
Rmat = eval.penalty(tempbasis, harmaccelLfd)
```

We hasten to add that most routine functional data analyses will not actually need to calculate roughness penalty matrices, since this happens inside functions such as smooth.basis. Computing $\mathbf{R}$ can involve numerical approximations to the integrals involved in (5.15). However, for a spline basis, if L is a power of D, then the integrals are analytically available and evaluated to within machine precision.

5.2.3 The Smoothing or "Hat" Matrix and Degrees of Freedom

The values $x(t_j), j = 1, \ldots, n$ defined by minimizing criterion (5.14) are critical for a detailed analysis of how well alternative choices λ work for fitting the data values y_j. Let us denote these by the vector $\hat{\mathbf{x}}$ and the corresponding data values by $\mathbf{y}$. It turns out (see Ramsay and Silverman (2005) for details) that $\hat{\mathbf{x}}$ has the following *linear* relationship to $\mathbf{y}$:

$$\hat{\mathbf{x}} = \mathbf{H}(\lambda)\mathbf{y}. \tag{5.19}$$

The *smoothing matrix* $\mathbf{H}(\lambda)$ is square, symmetric and of order n and, needless to say, a function of λ. It has many uses, among which is that a measure of the effective degrees of freedom of the fit defined by λ is defined by

$$df(\lambda) = trace[\mathbf{H}(\lambda)], \tag{5.20}$$

and the associated degrees of freedom for error is $n - df(\lambda)$.

As $\lambda \to 0, df(\lambda) \to \min(n, K)$, where $n =$ the number of observations and $K =$ the number of basis functions. Similarly, as $\lambda \to \infty, df(\lambda) \to m$, where m is the order of the highest derivative used to define the roughness penalty.

5.2.4 Defining Smoothing by Functional Parameter Objects

Going beyond the smoothing problem, we need the general capacity in functional data analysis to impose smoothness on estimated *functional parameters*, of which the smoothing curve is only one example. We now explain how this is made possible in the two programming languages.

A roughness penalty is defined by constructing a *functional parameter object* consisting of:

- a basis object,
- a derivative order m or a differential operator L to be penalized and
- a smoothing parameter λ.

We put these elements together by using the fdPar class in either language and the function fdPar to construct an object of that class.

The following R commands do two things: First they set up an order six B-spline basis for smoothing the growth data using a knot at each age. Then they define a functional parameter object that penalizes the roughness of growth acceleration by using the fourth derivative in the roughness penalty. The smoothing parameter value that we have found works well here is $\lambda = 0.01$.

```
norder = 6
nbasis = length(age) + norder - 2
heightbasis = create.bspline.basis(c(1,18),
                        nbasis, norder, age)
heightfdPar = fdPar(heightbasis, 4, 0.01)
```

The data are in array heightmat. In Chapter 4, these data were passed to smooth.basis with a basis object as the third argument. Here, we will use the functional parameter object heightfdPar as the third argument:

```
heightfd = smooth.basis(age, heightmat,
                        heightfdPar)$fd
```

Notice that we set up a functional data object heightfd directly by using the suffix $fd. In Matlab, we would use

```
heightfd = smooth_basis(age, heightmat, heightfdPar)
```

5.2.5 Choosing Smoothing Parameter λ

The *generalized cross-validation* measure GCV developed by Craven and Wahba (1979) is designed to locate a best value for smoothing parameter λ. The criterion is

$$\text{GCV}(\lambda) = \left(\frac{n}{n - df(\lambda)}\right)\left(\frac{\text{SSE}}{n - df(\lambda)}\right). \tag{5.21}$$

Notice that this is a twice-discounted mean square error measure. The right factor is the unbiased estimate of error variance σ^2 familiar in regression analysis, and thus represents some discounting by subtracting $df(\lambda)$ from n. The left factor further discounts this estimate by multiplying by $n/(n - df(\lambda))$.

Figure 5.1 shows how the generalized cross-validation (GCV) criterion varies as a function of $\log_{10}(\lambda)$ for the entire female Berkeley growth data. Matlab code for generating the plotted values is

```
loglam   = -6:0.25:0;
gcvsave = zeros(length(loglam),1);
dfsave  = gcvsave;
for i=1:length(loglam)
  lambdai   = 10^loglam(i);
  hgtfdPari = fdPar(heightbasis, 4, lambdai);
  [hgtfdi, dfi, gcvi] =
      smooth_basis(age, hgtfmat, hgtfdPari);
  gcvsave(i) = sum(gcvi);
  dfsave(i)  = dfi;
end
```

The minimizing value of λ is about 10^{-4}, and at that value $df(\lambda) = 20.2$. In fact, the value $\lambda = 10^{-4}$ is rather smaller than the value of 10^{-2} that we chose to work with in our definition of the fdPar object in Section 5.2.4, for which $df(\lambda) = 12.7$. We explain our decision in Section 5.3, and recommend a cautious and considered approach to choosing the smoothing parameter rather than relying solely on automatic methods such as GCV minimization.

GCV values often change slowly with $\log_{10}\lambda$ near the minimizing value, so that a fairly wide range of λ values may give roughly the same GCV value. This is a sign that the data are not especially informative about the "true" value of λ. If so, it is not worth investing a great deal of effort in precisely locating the minimizing value, and simply plotting GCV over a mesh of $\log_{10}\lambda$ might be sufficient. Plotting the function $GCV(\lambda)$ in any case will inform us about the curvature of near its minimum. If the data are not telling us all that much about λ, then it is surely reasonable to use your judgment in working with values which seem to provide more useful results than the minimizing value does. Indeed, Chaudhuri and Marron (1999) argue persuasively for inspecting data smooths over a range of λ values in order to see what is revealed at each level of smoothing. However, if a more precise value seems important, the function lambda2gcv can be used as an argument in an optimization function that will return the minimizing value.

5.3 Case Study: The Log Precipitation Data

The fda package for R includes CanadianWeather data, which includes the base 10 logarithms of the average annual precipitation in millimeters (after replacing

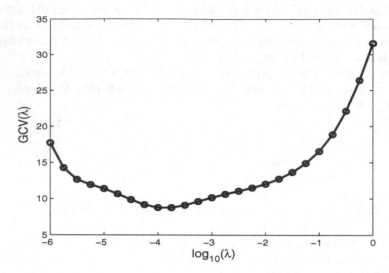

Fig. 5.1 The values of the generalized cross-validation or GCV criterion for choosing the smoothing parameter λ for fitting the female growth curves.

zeros with 0.05) for each day of the year at 35 different weather stations. We put these data in `logprecav`, shifted to put winter in the middle, so the year begins with July 1 and ends with June 30:

```
logprecav = CanadianWeather$dailyAv[
        dayOfYearShifted, ,'log10precip']
```

Next we set up a saturated Fourier basis for the data:

```
dayrange   = c(0,365)
daybasis   = create.fourier.basis(dayrange, 365)
```

We will smooth the data using a harmonic acceleration roughness penalty that penalizes departures from a shifted sine, $x(t) = c_1 + c_2 \sin(2\pi t/365) + c_3 \cos(2\pi t/365)$. Here we define this penalty. The first command sets up a vector containing the three coefficients required for the linear differential operator, and the second uses function `vec2Lfd` to convert this vector to the linear differential operator object `harmaccelLfd`.

```
Lcoef        = c(0,(2*pi/diff(dayrange))^2,0)
harmaccelLfd = vec2Lfd(Lcoef, dayrange)
```

Now that we are set up to do some smoothing, we will want to try a range of smoothing parameter λ values and examine the degrees of freedom and values of the generalized cross–validation coefficient GCV associated with each value of λ. First we set up a range of values (identified, of course, by some preliminary trial-

and-error experiments). We also set up two vectors to contain the degrees of freedom and GCV values.

```
loglam   = seq(4,9,0.25)
nlam     = length(loglam)
dfsave   = rep(NA,nlam)
gcvsave  = rep(NA,nlam)
```

Here are commands that loop through the smoothing values, storing degrees of freedom and GCV along the way:

```
for (ilam in 1:nlam) {
  cat(paste('log10 lambda =',loglam[ilam],'\n'))
  lambda   = 10^loglam[ilam]
  fdParobj = fdPar(daybasis, harmaccelLfd, lambda)
  smoothlist = smooth.basis(day.5, logprecav,
                            fdParobj)
  dfsave[ilam]  = smoothlist$df
  gcvsave[ilam] = sum(smoothlist$gcv)
}
```

The GCV values have to be summed, since function smooth.basis returns a vector of GCV values, one for each replicate.

Figure 5.2 plots the GCV values. This shows a minimum at $\log_{10}(\lambda) = 6$. Next we smooth at this level and add labels to the resulting functional data object. Then we plot all the log precipitation curves in a single plot, followed by a curve–by–curve plot of the raw data and the fitted curve.

```
lambda   = 1e6
fdParobj = fdPar(daybasis, harmaccelLfd, lambda)
logprec.fit = smooth.basis(day.5,logprecav,fdParobj)
logprec.fd = logprec.fit$fd
fdnames = list("Day (July 1 to June 30)",
               "Weather Station" = CanadianWeather$place,
               "Log 10 Precipitation (mm)")
logprec.fd$fdnames = fdnames
plot(logprec.fd)
plotfit.fd(logprecav, day.5, logprec.fd)
```

This example will be revisited in Chapter 7. There, we will see that the $\lambda =$ 1e6 leaves some interesting structure in the residuals for a few weather stations. Moreover, the curvature in the GCV function is rather weak, suggesting we will not lose much by using other values of λ in the range of 1e5 to 1e8. Our advice at the end of Section 5.2.5 seems appropriate here, and perhaps we should have worked with a lower value of λ.

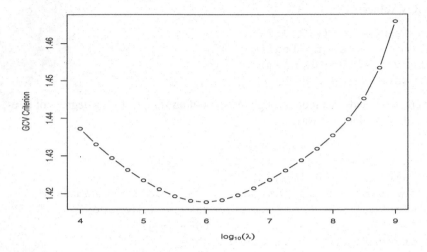

Fig. 5.2 The values of the generalized cross-validation or GCV criterion for the log precipitation data. The roughness penalty was defined by harmonic acceleration.

5.4 Positive, Monotone, Density and Other Constrained Functions

Often estimated curves must satisfy one or more side constraints. If the data are counts or other values that cannot be negative, then we do not want negative curve values, even over regions where values are at or close to zero. If we are estimating growth curves, it is probably the case that negative slopes are implausible, even if the noisy measurements do go down here and there. If the data are proportions, it would not make sense to have curve values outside the interval [0,1].

Unfortunately, linear combinations of basis functions such as those we have been using up to this point are difficult to constrain in these ways. The solution to the problem is simple: We transform the problem to one where the curve being estimated is unconstrained. We lose simple closed form expressions for the smoothing curve and therefore must resort to iterative methods for calculating the transformed curve, but the price is well worth paying.

5.4.1 Positive Smoothing

This transformation strategy is easiest to see in the case of positive (or negative) curves. We express the smoothing problem (5.3) as the transformed problem

$$y_j = \exp[w(t_j)] + \varepsilon_j = \exp[\phi(t_j)'\mathbf{c}] + \varepsilon. \tag{5.22}$$

That is, function $w(t)$ is now the logarithm of the data-fitting function $x(t) = \exp[w(t)]$, and consequently is unconstrained as to its sign, while at the same time the fitting function is guaranteed to be positive. It can go as close to zero as we like by permitting the values of $w(t)$ to be arbitrarily large negative numbers.

For example, we can smooth Vancouver's mean daily precipitation data, which can have zero but not negative values, using these commands using the function smooth.pos in R or smooth_pos in Matlab:

```
lambda    = 1e3
WfdParobj = fdPar(daybasis, harmaccelLfd, lambda)
VanPrec   = CanadianWeather$dailyAv[
   dayOfYearShifted, 'Vancouver', 'Precipitation.mm']
VanPrecPos = smooth.pos(day.5, VanPrec, WfdParobj)
Wfd = VanPrecPos$Wfdobj
```

These commands plot Wfd, the estimated log precipitation.

```
Wfd$fdnames = list("Day (July 1 to June 30)",
        "Weather Station" = CanadianWeather$place,
                "Log 10 Precipitation (mm)")
plot(Wfd)
```

The fit to the data, shown in Figure 5.3, is displayed by

```
precfit = exp(eval.fd(day.5, Wfd))
plot(day.5, VanPrec, type="p", cex=1.2,
    xlab="Day (July 1 to June 30)",
    ylab="Millimeters",
    main="Vancouver's Precipitation")
lines(day.5, precfit,lwd=2)
```

5.4.2 Monotone Smoothing

Some applications require a fitting function $x(t)$ that is either monotonically increasing or decreasing, even though the observations may not exhibit perfect monotonicity:

$$y_j = \beta_0 + \beta_1 x(t_j) + \varepsilon_j \tag{5.23}$$

We can get this easily by letting

$$x(t) = \int_{t_0}^{t} \exp[w(u)]\,du. \tag{5.24}$$

Here t_0 is the fixed origin for the range of t-values for which the data are being fit. The intercept term β_0 in (5.23) is the value of the approximating function at t_0.

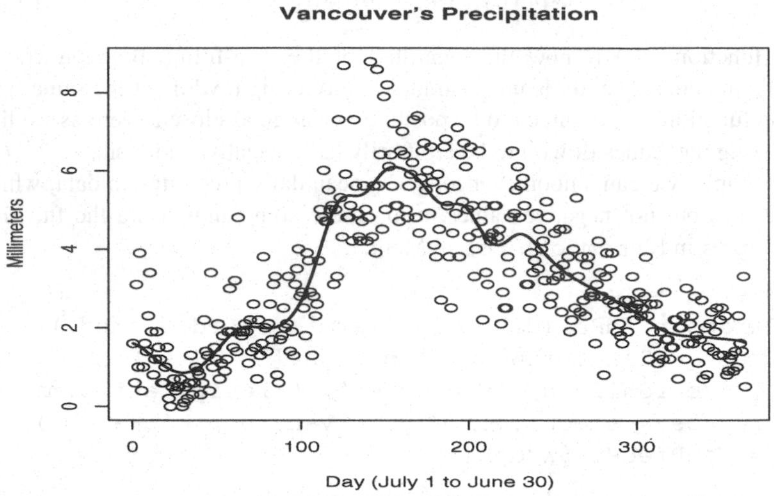

Fig. 5.3 Vancouver's precipitation data, along with a fit estimated by positive smoothing.

For monotonically increasing functions, β_1 could be absorbed into $w(u)$. However, to allow for monotonically decreasing functions, we keep β_1 separate and select normalize $w(u)$ for numerical stability.

Substituting (5.24) into (5.23) produces the following:

$$y_j = \beta_0 + \beta_1 \int_{t_0}^{t_j} \exp[w(u)]\,du + \varepsilon_j = \beta_0 + \beta_1 \int_{t_0}^{t_j} \exp[\phi(u)'\mathbf{c}]\,du + \varepsilon_j. \qquad (5.25)$$

The function `smooth.monotone` estimates β_0, β_1, and $w(u)$.

5.4.2.1 Smoothing the Length of a Newborn Baby's Tibia

Figure 5.4 shows the length of the tibia of a newborn infant, measured by Dr. Michael Hermanussen with an error of the order of 0.1 millimeters, over its first 40 days. The staircase nature of growth in this early period and need to estimate the velocity of change in bone length, also shown in the figure, makes monotone smoothing essential. It seems astonishing that this small bone in the baby's lower leg has the capacity to grow as much as two millimeters in a single day.

Variables `day` and `tib` in the following code contain the numbers of the days and the measurements, respectively. A basis for function w and a smoothing profile are set up, the data are smoothed, the values of the functional data object for w and the coefficients β_0 and β_1 are returned. Then the values of the smoothing and velocity curves are computed.

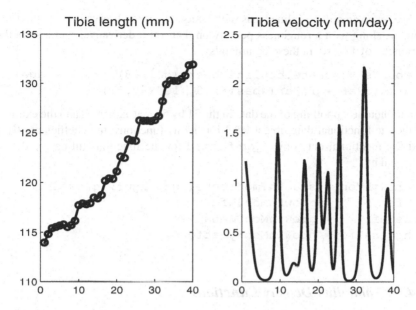

Fig. 5.4 The left panel shows measurements of the length of the tibia of a newborn infant over its first 40 days, along with a monotone smooth of these day. The right panel shows the velocity or first derivative of the smoothing function.

```
Wbasis  = create.bspline.basis(c(1,n), nbasis)
Wfd0    = fd(matrix(0,nbasis,1), Wbasis)
WfdPar  = fdPar(Wfd0, 2, 1e-4)
result  = smooth.monotone(day, tib, WfdPar)
Wfd     = result$Wfd
beta    = result$beta
dayfine = seq(1,n,len=151)
tibhat  = beta[1] + beta[2]*eval.monfd(dayfine ,Wfd)
Dtibhat = beta[2]*eval.monfd(dayfine, Wfd, 1)
D2tibhat = beta[2]*eval.monfd(dayfine, Wfd, 2)
```

5.4.2.2 Smoothing the Berkeley Female Growth Data

In Chapter 8 we will need our best estimates of the growth acceleration functions for the Berkeley girls, and smoothing their data monotonically substantially improves these estimates over direct smoothing, and especially in the neighborhood of the pubertal growth spurts.

We set up an order 6 spline basis with knots at ages of observations for their functions w, along with a roughness penalty on their third derivatives and a smoothing parameter of $1/\sqrt{10}$, in these commands:

```
wbasis = create.bspline.basis(c(1,18), 35, 6, age)
growfdPar = fdPar(wbasis, 3, 10^(-0.5))
```

The monotone smoothing of the data in the 31 by 54 matrix hgtf, and the extraction of the the functional data object Wfd for the w_i functions, the coefficients β_{0i}, β_{1i} and the functional data object hgtfhatfd for the functions fitting the data are achieved by

```
growthMon = smooth.monotone(age, hgtf, growfdPar)
Wfd       = growthMon$Wfd
betaf     = growthMon$beta
hgtfhatfd = growthMon$yhatfd
```

5.4.3 Probability Density Functions

A probability density function $p(z)$ is used to indicate the probability of observing a scalar observation at or near a value z, and is one of the core functions in statistics. From our perspective, p is a special case of a positive function in the sense of having a unit integral over the range of z. That is, we can express p as

$$p(z) = C \exp[w(z)], \tag{5.26}$$

where the positive *normalizing constant* C satisfies the constraint $\int p(z)\mathrm{d}z = 1$. Estimating a free-form nonparametric version of p is not a smoothing problem as we have so far defined it, since we would not use an error sum of squares measure of lack of fit. Rather, the usual practice would be to minimize a *penalized negative log likelihood*,

$$-\ln L(\mathbf{c}) = -\sum_i^N \ln p(z_i) + \lambda \int [Lw(z)]^2 \mathrm{d}z = -\sum_i^N w(z_i) - N\ln C + \lambda \int [Lw(z)]^2 \mathrm{d}z, \tag{5.27}$$

where $w(z) = \mathbf{c}'\phi(z)$. Notice that the first two terms replace the error sum of squares in (5.13).

The linear differential operator L can be chosen so as to force p to approach specific parametric density functions as $\lambda \to \infty$. For example, $L = D^3$ will do this for the Gaussian density (Silverman, 1986).

Function density.fd is used to estimate a nonparametric probability density function from a sample of data. We will illustrate its use for a rather challenging problem: describing the variation in daily precipitation for the Canadian prairie city of Regina over the month of June and over the 34 years from 1960 through 1993. June is the critical month for wheat growers because the crop enters its most rapid growing phase, and an adequate supply of moisture in the soil is essential for a good crop.

Precipitation is a difficult quantity to model for several reasons. First of all, on about 65% of the days in this region, no rain is even possible, so that zero really means a "nonprecipitation day" rather than "no rain." Since there can be a small amount of precipitation from dew, we used only days when the measured precipitation exceeded two millimetres. Also, precipitation can come down in two main ways: as a gentle drizzle and, more often, as a sudden and sometimes violent thunderstorm. Consequently, the distribution of precipitation is extremely skewed, and Regina experienced three days in this period with more than 40 mm of rain. We deleted these days, too, in order to improve the graphical displays, leaving $N = 212$ rainfall values.

Figure 5.5 plots the *ordered* rainfalls for the 1,006 days when precipitation was recorded against their rank orders, a version of a *quantile plot*. We can see just how extreme precipitation can be; the highest rainfall of 132.6 mm on June 25, 1975, is said to have flooded 20,000 basements.

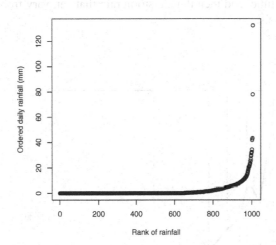

Fig. 5.5 The empirical quantile function for daily rainfall at Regina in the month of June over 34 years.

We set up the break points for a cubic B-spline basis to be the rainfalls at 11 equally spaced ranks, beginning at the first and ending at N. In this code variable RegPrec contains the 212 sorted rainfall amounts between 2 and 45 mm.

```
Wknots  = RegPrec[round(N*seq(1/N,1,len=11),0)]
Wnbasis = length(Wknots) + 2
Wbasis  = create.bspline.basis(range(RegPrec),13,4,
```

Wknots)

Now we estimate the density, applying a light amount of smoothing, and extract the functional data object Wfd and the normalizing constant C from the list that density.fd returns.

```
Wlambda       = 1e-1
WfdPar        = fdPar(Wbasis, 2, Wlambda)
densityList   = density.fd(RegPrec, WfdPar)
Wfd           = densityList$Wfdobj
C             = densityList$C
```

These commands set up the density function values over a fine mesh of values.

```
Zfine = seq(RegPrec[1],RegPrec[N],len=201)
Wfine = eval.fd(Zfine, Wfd)
Pfine = exp(Wfine)/C
```

The estimated density is shown in Figure 5.6. The multiphase nature of precipitation is clear here. The first phase is due to heavy dew or a few drops of rain, followed by a peak related to light rain from low-pressure ridges that arrive in this area from time to time, and then thunderstorm rain that can vary from about 7 mm to catastrophic levels.

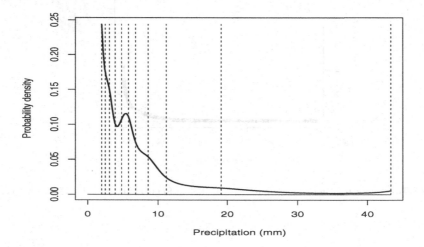

Fig. 5.6 The solid line indicates the probability density function $p(z)$ for rainfall in Regina of 2 mm or greater, but stopping at about 45 mm. The vertical dashed lines indicate the knot values used to define the cubic B-spline expansion for $w = \ln p$.

5.5 Assessing the Fit to the Data

Having smoothed the data, there are many questions to ask, and these direct us to do some further analyses of the residuals, $r_{ij} = y_{ij} - x_i(t_j)$. These analyses can be functional since there is some reason to suppose that at least part of the variation in these residuals across t is smooth.

Did we miss some important features in the data by oversmoothing? Perhaps, for example, there may have been something unusual in one or two curves that we missed because the GCV criterion selected a level of smoothing that worked best for all samples simultaneously. Put another way, could there be an indication that we might have done better to smooth each weather station's log precipitation data separately? We will defer looking at this question until the end of the next chapter, since principal components analysis can be helpful here.

A closely related question concerns whether the variation in the residuals conforms to the assumptions implicit in the type of smoothing that we performed. The use of the unweighted least-squares criterion is only optimal if the residuals for all time points are normally distributed and if the variance of these residuals is constant across both years and weather stations (curves).

We now return to the log precipitation data considered in Section 5.3 and create a 365 by 35 matrix of residuals from the fit discussed there. We then use this to create variance vectors across

- stations, of length 365, dividing by 35 since the residuals need not sum to zero on any day,
- time, of length 35, dividing by 365-12; the number "12" here is essentially the equivalent degrees of freedom in the fit (logprec.fit$df).

```
logprecmat = eval.fd(day.5, logprec.fd)
logprecres = logprecav - logprecmat
#  across stations
logprecvar1 = apply(logprecres^2, 1, sum)/35
#  across time
logprecvar2 = apply(logprecres^2, 2, sum)/(365-12)
```

Let us look at how residual variation changes over stations; Figure 5.7 displays their standard deviations. With labels on a few well-known stations and recalling that we number the stations from east to west to north, we see that there tends to be more variation for prairie and northerly stations in the center of the country, and less for marine stations. This is interesting but perhaps not dramatic enough to make us want to pursue the matter further.

Figure 5.8 shows how standard deviations taken over stations and within days vary. The smooth line in the plot was computed by smoothing the log of the standard deviations and exponentiating the result by these two commands:

```
logstddev.fd = smooth.basis(day.5,
                     log(logprecvar1)/2, fdParobj)$fd
logprecvar1fit = exp(eval.fd(day.5, logstddev.fd))
```

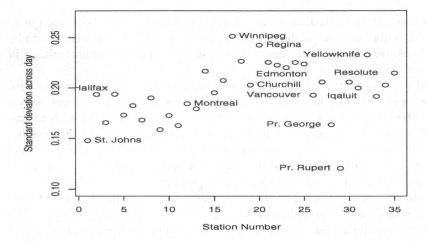

Fig. 5.7 Standard deviations of the residuals from the smooth of the log precipitation taken across days and within stations.

We could also have used `smooth.pos` to do the job. We see now that there is a seasonal variation in the size of the residuals, with more variation in summer months than in winter. Nevertheless, this form of variation is not strong enough to justify returning to do a weighted least-squares analysis using `smooth.basis`; we would need much larger variations in the variability for it to create a substantive difference between weighted and unweighted solutions.

Also implicit in our smoothing technology is the assumption that residuals are uncorrelated. This is a rather unlikely situation; departures from smooth variation tend also to be smooth, implying a strong positive autocorrelation between neighboring residuals. If observation times are equally spaced, we can use standard time series techniques to explore this autocorrelation structure.

5.6 Details for the `fdPar` Class and `smooth.basis` Function

5.6.1 The `fdPar` class

We give here the arguments of the constructor function `fdPar` that constructs an object of the functional parameter `fdPar` class. The complete calling sequence is

```
fdPar(fdobj=NULL, Lfdobj=NULL, lambda=0,
      estimate=TRUE, penmat=NULL)
```

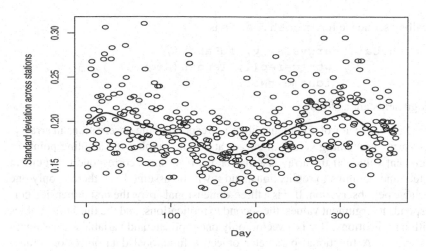

Fig. 5.8 Standard deviations of the residuals from the smooth of the log precipitation taken across stations and within days. The solid line is an exponentiated smooth of the log of the variances.

The arguments are as follows:

`fdobj`   A functional data object, functional basis object, a functional parameter object or a matrix. If it a matrix, it is replaced by `fd(fdobj)`. If `class(fdobj) == 'basisfd'`, it is converted to an object of class `fd` with a coefficient matrix consisting of a single column of zeros.

`Lfdobj`   Either a nonnegative integer or a linear differential operator object. If NULL, Lfdobj depends on `fdobj[['basis']][['type']]`:

bspline   Lfdobj = `int2Lfd(max(0, norder-2))`, where norder = `norder(fdobj)`.

fourier   Lfdobj is a harmonic acceleration operator set up for the period used to define the basis.

anything else   Lfdobj <- `int2Lfd(0)`

`lambda`   A nonnegative real number specifying the amount of smoothing to be applied to the estimated functional parameter.

`estimate`   Not currently used.

`penmat`   A roughness penalty matrix. Including this can eliminate the need to compute this matrix over and over again in some types of calculations.

5.6.2 The `smooth.basis` Function

The calling sequence for `smooth.basis` is

```
smooth.basis(argvals, y, fdParobj,
             wtvec=rep(1, length(argvals)),
             fdnames=NULL)
```

The arguments are as follows:

argvals   A vector of argument values correspond to the observations in array y.

y   An array containing values of curves at a finite number of sampling points or argument values. If the array is a matrix, the rows must correspond to argument values and columns to replications, and it will be assumed that there is only one variable per observation. If y is a three-dimensional array, the first dimension corresponds to argument values, the second to replications, and the third to variables within replications. If y is a vector, only one replicate and variable are assumed.

fdParobj   A functional parameter object, a functional data object or a functional basis object. If the object is a functional parameter object, then the linear differential operator object and the smoothing parameter in this object define the roughness penalty. If the object is a functional data object, the basis within this object is used without a roughness penalty, and this is also the case if the object is a functional basis object.

wtvec   A vector of the same length as argvals containing weights for the values to be smoothed.

fdnames   A list of length three containing character vectors of names for the following:

args   Names for each observation or point in time at which data are collected.

reps   Names for each rep, unit or subject.

fun   Names for each function or (response) variable measured repeatedly (per args) for each rep.

Function `smooth.basis` returns an argument of the fdSmooth class, which is a named list of length eight with the following components:

fd   A functional data object containing a smooth of the data.

df   A degrees of freedom measure of the smooth.

gcv   The value of the generalized cross-validation or GCV criterion. If there are multiple curves, this is a vector of values, one per curve. If the smooth is multivariate, the result is a matrix of GCV values, with columns corresponding to variables.

SSE   The error sums of squares; SSE is a vector or a matrix of the same size as gcv.

penmat   The penalty matrix.

y2cMap   The matrix mapping the data to the coefficients.

argvals, y   Input arguments

5.7 Some Things to Try

1. In order to understand the implications of choice of smoothing parameter λ, it is best to work with simulated data.

   a. Choose a function with some interesting variation, such as
      - $\sin(4\pi t)$ over [0,1]
      - $\exp(-t^2/2)$ over [-5,5]
   b. Specifying some sampling points $t_j, j = 1, \ldots, n$, evaluate your function at these points, and add some mean 0 normal random error to these values, where you specify the standard deviation.
   c. Specify a basis system, an order of derivative to penalize, and a value smoothing parameter λ, and bundle these together in an `fdPar` object.
   d. Smooth the points using function `smooth.basis` (R) or `smooth_basis` (Matlab).
   e. Compute the square root of the mean square error, and compare this to the standard deviation that you used to generate the error. You might want to subtract the equivalent degrees of freedom associated with the λ value that you used from n before dividing the sum of squared errors.
   f. Experiment with different values of λ, and find one that gives nearly the right value for root mean square error.
   g. Experiment with different values of λ, and save the GCV values for each. Plot these GCV values against $\log_{10}(\lambda)$, and estimate by eye the value of λ minimizing GCV.

2. Now try your hand at smoothing some real data, such as what you used in the exercises in Chapter 1.
3. Calculate the derivative of the function you used to generate the data in 1a. How does the derivative of your smooth compare with this? Does some value for λ other than that given by GCV improve the agreement between estimated and true derivatives?
4. **Melanoma Data** The `melanoma` data set in the `fda` package for R contains age-adjusted incidences of melanoma from the Connecticut Tumor Registry for the years 1935 to 1972.

   a. Fit these data with a Fourier basis, choosing the number of basis functions by minimizing the `gcv` value returned by `smooth.basis`.
   b. Try removing a linear trend for these data first, either directly or by looking at the residuals after a call to `lm`. Repeat the steps above; does the optimal number of basis functions change?
   c. Refit the data using a B-spline basis and a harmonic acceleration penalty. Try some values of λ to optimize `gcv`. You will need to guess at the period to use; how does doubling and halving the period change the degrees of freedom at the optimal value of λ?
   d. Set up the linear differential operator $\omega^2 D^2 + D^4$, which annihilates sinusoidal combined with linear trend. Smooth the melanoma data using this operator.

e. Plot the velocity versus acceleration curves for the fit using a Fourier basis
and using the B-spline basis with a harmonic acceleration penalty. Are they
substantially different? Do they provide evidence of subcycles?

5.8 More to Read

There is a large literature on smoothing methods, and Ramsay and Silverman (2005)
devote a number of chapters to the problem. Recent book-length references are Eubank (1999), Rupert et al. (2003), and Simonoff (1996). Moreover, there are smoothing methods that do not define x explicitly in terms of basis functions that may serve
as well, such as *local polynomial smoothing*. However, the well-known method
of *kernel smoothing*, made all too available in software packages, should now be
viewed as obsolete because its poor performance near the end points of the interval
(Fan and Gijbels, 1996).

Chapter 6
Descriptions of Functional Data

This chapter and the next are the exploratory data analysis end of functional data analysis. Here we recast the concepts of mean, standard deviation, covariance and correlation into functional terms and provide R and Matlab functions for computing and viewing them.

Exploratory tools are often the most fruitful when applied to *residual variation* around some model, where we often see surprising effects once we have removed relatively predictable structures from the data. Summary descriptions of residual variation are also essential for estimating confidence regions.

Contrasts are often used in analysis of variance to explore prespecified patterns of variation. We introduce the more general concept of a *functional probe* as a means of looking for specific patterns or shapes of variation in functional data and of providing methods for estimating confidence limits for estimated probe values.

The phase-plane plot has turned out to be a powerful tool for exploring data for harmonic variation, even in data on processes such as human growth where we do not ordinarily think of cyclic variation as of much interest. It is essentially a graphical analogue of a second order linear differential equation. In fact, the phase-plane plot, developed in detail in this chapter, is a precursor to the dynamic equations that we will explore in Chapter 11.

6.1 Some Functional Descriptive Statistics

Let $x_i, i = 1, \ldots, N$, be a sample of curves or functions fit to data. The univariate summaries, the sample mean and variance functions, are as follows:

$$\bar{x}(t) = N^{-1} \sum_i x_i(t) \quad \text{and} \quad s(t) = (N-1)^{-1} \sum_i [x_i(t) - \bar{x}(t)]^2.$$

These are computed, for the log-precipitation data considered in Section 5.3, as follows:

J.O. Ramsay et al., *Functional Data Analysis with R and MATLAB*, Use R,
DOI: 10.1007/978-0-387-98185-7_6,
© Springer Science + Business Media, LLC 2009

```
meanlogprec    = mean(logprec.fd)
stddevlogprec  = std.fd(logprec.fd)
```

As always in statistics, choices of descriptive measures like the mean and variance should never be automatic or uncritical. The distribution of precipitation is strongly skewed, and by logging these data, we effectively work with the geometric mean of precipitation as a more appropriate measure of location in the presence of substantial skewness.

Beyond this specific application, the functional standard deviation focuses on the intrinsic variability between observations, e.g., Canadian weather stations, after removing variations that are believed to represent measurement and replication error not attributable to the variability between observations. A proper interpretation of the analyses of this section require an understanding of exactly what we mean by std.fd and what is discarded in smoothing.

6.1.1 The Bivariate Covariance Function $v(s,t)$

As we indicated in Chapter 1, the correlation coefficient as a measure of association between two functional observations $x_i(s)$ and $x_i(t)$ on the same quantity or *metric* is often less useful than the simpler covariance coefficient, because they share the same measurement scales. Where we want to quantify the association between two functions x and y having different measurement scales, the correlation will still be useful.

The bivariate covariance function $\sigma(s,t)$ specifies the *covariance* between curve values $x_i(s)$ and $x_i(t)$ at times s and t, respectively. It is estimated by

$$v(s,t) = (N-1)^{-1} \sum_i [x_i(s) - \bar{x}(s)][x_i(t) - \bar{x}(t)]. \qquad (6.1)$$

For the log-precipitation data the R command is

```
logprecvar.bifd = var.fd(logprec.fd)
```

The result of this command is a bivariate functional data object having two arguments. If we want to look at the *variance-covariance surface*, these commands in Matlab will do the job:

```
logprecvar_bifd = var_fd(logprec_fd);
weektime        = linspace(0,365,53);
logprecvar_mat  = eval_bifd(weektime, weektime,
                             logprecvar_bifd);
surf(weektime, weektime, logprecvar_mat);
contour(weektime, weektime, logprecvar_mat);
```

The following will do essentially the same thing in R:

```
weektime         = seq(0,365,length=53)
```

```
logprecvar_mat   = eval.bifd(weektime, weektime,
                                  logprecvar.bifd)
persp(weektime, weektime, logprecvar_mat,
     theta=-45, phi=25, r=3, expand = 0.5,
     ticktype='detailed',
     xlab="Day (July 1 to June 30)",
     ylab="Day (July 1 to June 30)",
     zlab="variance(log10 precip)")
contour(weektime, weektime, logprecvar_mat)
```

In our experience, contour and three-dimensional surface or perspective plots com-
plement each other in the information that they convey, and both are worth doing.
Surface plots draw our eye to global shape features, but we need contour plots to
locate these features on the argument plane.

Function `var.fd` may also be used to compute the `cross-covariance` between two sets of curves by being called with two arguments, such as in

```
tempprecbifd = var.fd(tempfd, logprec.fd)
```

If the *cross-correlation surface* is needed, however, we would use the function
`cor.fd` or its Matlab counterpart `cor_fd`.

The variance of the log precipitation functions is seen in Figure 6.1 as the height
of the diagonal running from (0,0) to (365,365). There is much more variation in
precipitation in the winter months, positioned in this plot in the middle of the surface, because the frigid atmosphere near polar stations like Resolute has almost no
capacity to carry moisture, while marine stations like Prince Rupert are good for a
soaking all year round. One is struck by the topographical simplicity of this particular surface, and we will understand this better in the next section.

The R commands

```
day5time = seq(0,365,5)
logprec.varmat = eval.bifd(day5time, day5time,
                      logprecvar.bifd)
contour(day5time, day5time, logprec.varmat,
        xlab="Day (July 1 to June 30)",
        ylab="Day (July 1 to June 30)", lwd=2,
        labcex=1)
```

return the contour plot of the variance surface shown in Figure 6.2. We see that
variance across weather stations is about five times as large in the winter than it is
in the summer. The action is in winter in Canada!

The documentation for the `surf` and `contour` functions in Matlab describe
enhancements over the images visible in Figures 6.1 and 6.2. With R, other perspective and contour functions are available in the `lattice` (Sarkar, 2008) and `rgl`
(Adler and Murcoch, 2009) packages. In particular, the `lattice` package is useful for high-dimensional graphics, showing, e.g., how the relationships displayed
in Figures 6.1 and 6.2 vary with region of the country. The `rgl` package provides
interactive control over perspective plots.

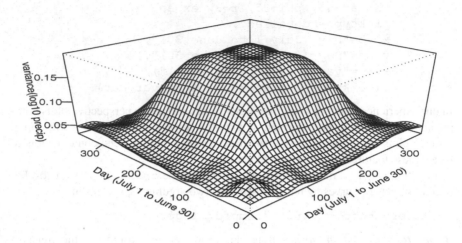

Fig. 6.1 The estimated variance-covariance surface $v(s,t)$ for the log precipitation data.

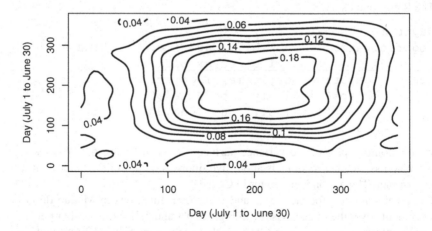

Fig. 6.2 A contour plot of the bivariate variance-covariance surface for the log precipitation data.

6.2 The Residual Variance-Covariance Matrix Σ_e

We considered the question of how the residuals $r_{ij} = y_{ij} - x_i(t_j)$ behave in Section 5.5, and we will return to this question in Chapters 7 and 8. But in the meantime we will need the *conditional covariance matrix* or *residual covariance matrix* describing the covariance of the residuals r_{ij} at argvals t_j, $j = 1, ..., n$. This is an order n symmetric matrix Σ_e. Here the term *conditional* means the variation in the y_{ij}'s left unexplained by a smooth curve or by the use of some other model for the data. We will use this matrix for computing confidence limits for curves and other values.

Unless a large number of replications of curves are available, as is the case for the growth data, we have to restrict our aims to estimating fairly gross structure in the residuals. In particular, it is often assumed that neighboring residuals are uncorrelated, and one only attempts to estimate the standard deviation or variance of the residuals across curves. Figure 5.8 offers a picture of this variation for the log precipitation data. Under this assumption, the order n symmetric matrix Σ_e will be diagonal and will contain values in the vector `logprecvar1` computed in Section 5.5.

We will consider ways of extracting more information Σ_e in Chapter 7.

6.3 Functional Probes ρ_ξ

Purely descriptive methods such as displaying mean and variance functions allow us to survey functional variation without having to bring any preconceptions about exactly what kind of variation might be important. This is fine as far as it goes, but functions and their derivatives are potentially complex structures with a huge scope for surprises, and we may need to "zoom in" on certain curve features.

Moreover, our experience suggests that a researcher seldom approaches functional data without some fairly developed sense of what will be seen. We would be surprised if we did not see the pubertal growth spurt in growth curves or sinusoidal variation in temperature profiles. When we have such a structure in mind, we typically need to do two things: check the data to be sure that what we expect is really there, and then do something clever to look around and beyond what we expect in order to view the unexpected. Chapter 7 is mainly about looking for the dominant modes of variation and covariation, but the tools that we develop there can also be used to highlight interesting but more subtle features.

A *probe* ρ_ξ is a tool for highlighting specific variation. Probes are variably weighted linear combinations of function values. Let ξ be a weight function that we apply to a function x as follows:

$$\rho_\xi(x) = \int \xi(t)x(t)\,dt. \tag{6.2}$$

If ξ has been structured so as to be a template for a specific feature or pattern of variation in x, then the resulting probe value $\rho_\xi(x)$ will be substantially far from zero. The term *contrast* in experimental design or linear models has much the same meaning as probe, but there is no particular need for probe functions to integrate to zero.

The value of a probe function is computing using the `inprod` function. Suppose `xifd` and `xfd` are two functional data objects for the weight function ξ and observed curve x, respectively. The probe value `probeval` is computed by the command

```
probeval = inprod(xifd, xfd)
```

The integration in this calculation can be done to within machine precision in many cases, or otherwise is computed by a numerical approximation method.

Probe weight functions ξ may also be estimated from the data rather than chosen a priori. Two methods discussed in Chapter 7, *principal components analysis* and *canonical correlation analysis*, are designed to estimate probes empirically that highlight large sources of variation or covariation.

6.4 Phase-Plane Plots of Periodic Effects

The two concepts of energy and of functional data having variation on more than one timescale lead to the graphical technique of plotting one derivative against another, something that we will call *phase-plane plotting*. We saw an example in Figure 1.15 for displaying the dynamics in human growth.

We now return to the US nondurable goods manufacturing index, plotted in Figures 1.3 and 1.4, to illustrate these ideas. A closer look at a comparatively stable period, 1964 to 1967 shown in Figure 6.3, suggests that the index varies fairly smoothly and regularly within each year. The solid line is a smooth of these data using the roughness penalty method described in Chapter 5. We now see that the variation within this year is more complex than Figure 1.4 can possibly reveal. This curve oscillates three times during the year, with the size of the oscillation being smallest in spring, larger in the summer, and largest in the autumn. In fact, each year shows smooth variation with a similar amount of detail, and we now consider how we can explore these within-year patterns.

6.4.1 Phase-Plane Plots Show Energy Transfer

Now that we have derivatives at our disposal, we can learn new things by studying how derivatives relate to each other. Our tool is the plot of acceleration against velocity. To see how this might be useful, consider the phase-plane plot of the function $\sin(2\pi t)$, shown in Figure 6.4. This simple function describes a basic *harmonic pro-*

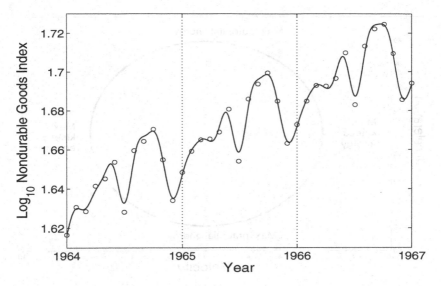

Fig. 6.3 The log nondurable goods index for 1964 to 1967, a period of comparative stability. The solid line is a fit to the data using a polynomial smoothing spline. The circles indicate the value of the log index at the first of the month.

cess, such as the vertical position of the end of a suspended spring bouncing with a period of one time unit.

Springs and pendulums oscillate because energy is exchanged between two states: *potential* and *kinetic.* At times $\pi, 3\pi, \ldots$ the spring is at one or the other end of its trajectory, and the restorative force due to its stretching has brought it to a standstill. At that point, its potential energy is maximized, and so is the force, which is acting either upward (positively) or downward. Since force is proportional to acceleration, the second derivative of the spring position, $-(2\pi)^2 \sin(2\pi t)$, is also at its highest absolute value, in this case about ± 40. On the other hand, when the spring is passing through the position 0, its velocity, $2\pi \cos(2\pi t)$, is at its greatest, about ± 8, but its acceleration is zero. Since kinetic energy is proportional to the square of velocity, this is the point of highest kinetic energy. The phase-plane plot shows this energy exchange nicely, with potential energy being maximized at the extremes of Y and kinetic energy at the extremes of X.

The amount of energy in the system is related to the width and height of the ellipse in Figure 6.4; the larger it is, the more energy the system exhibits, whether in potential or kinetic form.

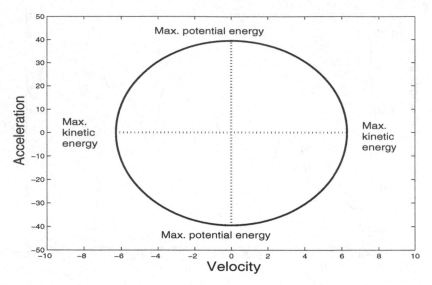

Fig. 6.4 A phase-plane plot of the simple harmonic function $\sin(2\pi t)$. Kinetic energy is maximized when acceleration is 0, and potential energy is maximized when velocity is 0.

6.4.2 The Nondurable Goods Cycles

Harmonic processes and energy exchange are found in many situations besides mechanics. In economics, potential energy corresponds to resources including capital, human resources, and raw material that are available to bring about some economic activity. This energy exchange can be evaluated for nondurable goods manufacturing as displayed in Figure 6.3. Kinetic energy corresponds to the manufacturing process in full swing, when these resources are moving along the assembly line and the goods are being shipped out the factory door.

We use the phase-plane plot, therefore, to study the energy transfer within the economic system. We can examine the cycle within individual years, and also see more clearly how the structure of the transfer has changed throughout the 20th century. Figure 6.5 presents a phase-plane plot for 1964, a year in a relatively stable period for the index. To read the plot, find "jan" in the middle right of the plot and move around the diagram clockwise, noting the letters indicating the months as you go. You will see that there are two large cycles surrounding zero, plus some small cycles that are much closer to the origin.

The largest cycle begins in mid-May (M), with positive velocity and near zero acceleration. Production is increasing linearly or steadily at this point. The cycle moves clockwise through June ("Jun") and passes the horizontal zero acceleration line at the end of the month, when production is now decreasing linearly. By mid-July ("Jly") kinetic energy or velocity is near zero because vacation season is in full swing. But potential energy or acceleration is high, and production returns to the

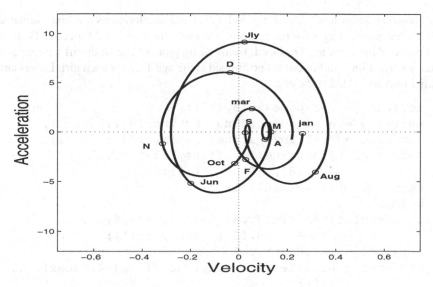

Fig. 6.5 A phase-plane plot of the first derivative or velocity and the second derivative or acceleration of the smoothed log nondurable goods index for 1964. Midmonths are indicated by the first letters or short abbreviations.

positive kinetic/zero potential phase in early August ("Aug"), and finally concludes with a cusp at summer's end (S). At this point the process looks like it has run out of both potential and kinetic energy.

The cusp, near where both derivatives are zero, corresponds to the start of school in September and the beginning of the next big production cycle passing through the autumn months of October through November. Again this large cycle terminates in a small cycle with little potential and kinetic energy. This takes up the months of February and March (F and mar). The tiny subcycle during April and May seems to be due to the spring holidays, since the summer and fall cycles, as well as the cusp, do not change much over the next two years, but the spring cycle cusp moves around, reflecting the variability in the timings of Easter and Passover.

To summarize, the production year in the 1960s has two large cycles swinging widely around zero, each terminating in a small cusplike cycle. This suggests that each large cycle is like a balloon that runs out of air, the first at the beginning of school and the second at the end of winter. At the end of each cycle, it may be that new resources must be marshaled before the next production cycle can begin.

6.4.3 Phase-Plane Plotting the Growth of Girls

Here are the commands in Matlab used to produce Figure 1.15. They use a functional data object `hgtfmonfd` that contains the 54 curves for the Berkeley girls estimated by monotone smoothing. Velocities and accelerations are first evaluated over a fine mesh of ages for the first ten girls using the `eval_fd` function. Then all 10 phase-plane plots are produced, followed by plots of the sixth girl's curve as a heavy dashed line, and of circles positioned at the age 11.7 for each girl. Labels and axis limits are added at the end.

```
agefine   = linspace(1,18,101);
velffine = eval_fd(agefine, hgtfmonfd(1:10), 1);
accffine = eval_fd(agefine, hgtfmonfd(1:10), 2);
phdl = plot(velffine, accffine, 'k-', ...
            [1,18], [0,0], 'k:');
set(phdl, 'LineWidth', 1)
hold on
phdl = plot(velffine(:,6), accffine(:,6), ...
            'k--', [0,12], [0,0], 'k:');
set(phdl, 'LineWidth', 2)
phdl=plot(velffine(64,index), accffine(64,index), ...
          'ko');
set(phdl, 'LineWidth', 2)
hold off
xlabel('\fontsize{13} Velocity (cm/yr)')
ylabel('\fontsize{13} Acceleration (cm/yr^2)')
axis([0,12,-5,2])
```

What we see is that girls with early pubertal growth spurts, having marker circles near the end of their trajectories, have intense spurts, indicated by the size of their loops. Late-spurt girls have tiny loops. The net effect is that the adult height of girls is not much affected by the timing of the growth spurt, since girls with late spurts have the advantage of a couple of extra years of growth, but the disadvantage of a weak spurt. A hint of the complexity of growth dynamics in infancy is given by the two girls whose curves come from the right rather than from the bottom of the plot.

6.5 Confidence Intervals for Curves and Their Derivatives

indexderivatives!confidence intervals We now want to see how to compute confidence limits on some useful quantities that depend on an estimated function x that has, in turn, been computed by smoothing with a roughness penalty a data vector $\mathbf{y}$. For example, how precisely is the function value at t, $x(t)$, determined by our sample of data $\mathbf{y}$? Or, what sampling standard deviation can we expect if we re-sample the data over and over again, estimating $x(t)$ anew with each sample? Can we construct a pair of *confidence limits* such that the probability that the true value of $x(t)$ lies within these limits is a specified value, such as 0.95? Displaying functions or their derivatives with pointwise confidence limits is a useful way of conveying how much information there is in the data used to estimate these functions. See Figure 6.6 below for an example.

More generally, confidence regions are often required for the values of linear probes ρ_ξ defined in (6.2), of which $x(t)$ and $D^m x(t)$ are specific examples.

6.5.1 Two Linear Mappings Defining a Probe Value

In order to study the sampling behavior of ρ_ξ, we need to compute two linear mappings plus their composite. They are given names and described as follows:

1. Mapping y2cMap, which converts the raw data vector **y** to the coefficient vector **c** of the basis function expansion of x. If **y** and **c** have lengths n and K, respectively, this mapping is a K by n matrix y2cMap such that

$$\mathbf{c} = \text{y2cMap}\, \mathbf{y}$$

where the K by n matrix y2cMap was defined in Chapter 5 by either (5.5) or (5.17).

2. Mapping c2rMap, which converts the coefficient vector **c** to the scalar quantity $\rho_\xi(x)$. This mapping is a 1 by K row vector **L** such that

$$\rho_\xi(x) = \mathbf{L}\mathbf{c} = \text{c2rMap}\, \mathbf{c}.$$

3. The composite mapping called y2rMap defined by

$$\text{y2rMap} = \rho_\xi(x) = \text{c2rMap}\ \text{y2cMap},$$

which converts a data vector **y** directly into the probe value; this is a 1 by n row vector.

How is $\mathbf{L} = \text{c2rMap}$ actually calculated? In general, the computation includes the use of the all-important *inner product function* inprod to compute the integral (6.2). This function is working away behind the scenes in almost every functional data analysis. It evaluates the integral of the product of two functions (or the matrices defined by products of sets of functions), such as that defining the roughness penalty matrix $\mathbf{R} = \int L\phi L\phi'$ defined in Subsection 5.2.2. Where possible, this function uses an analytic expression for these integral values. However, more often than not, this computation requires numerical approximation.

The four important arguments to function inprod are as follows:

fdobj1   Either a functional data object or a functional basis object.

fdobj2   Also either a functional data object or a functional basis object. It is the integral of the products of these two objects that is computed. If either of these first two arguments are a basis object, it is converted to a functional data object with an identity matrix as its coefficient matrix.

Lfdobj1   A linear differential operator object of class Lfd to be applied to
    fdobj1. If missing, the result of applying it is taken to be the function itself,
    that is, it is the *identity operator*.
Lfdobj2   Also a linear differential operator object of class Lfd to be applied to
    fdobj2.

For the problem of computing c2rMap, one of the first two arguments would
be a functional data object for the weight function ξ; the other would be the functional basis object used in the expansion of function x. As an illustration, consider a
conventional linear regression model with design matrix $\mathbf{Z}$

$$\mathbf{y} = \mathbf{Z}\mathbf{c} + \mathbf{e},$$

where the regression coefficient vector $\mathbf{c}$ is estimated by ordinary least squares.
Then, since $\mathbf{c} = (\mathbf{Z}'\mathbf{Z})^{-1}\mathbf{Z}'\mathbf{y}$, the matrix corresponding to y2cMap is $\mathbf{S} = (\mathbf{Z}'\mathbf{Z})^{-1}\mathbf{Z}'$.
Now suppose that for some reason we want to estimate the difference between the
first and second regression coefficients, possibly because we conjecture that they
may be equal in the population. Then the probe function ξ is equivalent to the
probe vector $\mathbf{L} = (1, -1, 0, \ldots)$, and this is the row vector corresponding to mapping c2rMap. Finally, the composite mapping y2rMap taking $\mathbf{y}$ directly into the
value of this difference is simply the row vector $\mathbf{L}(\mathbf{Z}'\mathbf{Z})^{-1}\mathbf{Z}'$.

For a more complicated example, suppose that we want to compare winter temperatures and precipitations for the 35 Canadian weather stations, and we have already defined basis objects tempbasis and precbasis, respectively. Suppose,
too, that we have run the year from July 1 to June 30, so that winter is in the middle
of the year. We can use as a probe function

$$\xi(t) = \exp\{20\cos[2 * \pi(t - 197)/365]\},$$

which is proportional to the density for the von Mises distribution of data on a
circle; the concentration parameter value 20 weights substantially about two months,
and the location value 197 centers the weighting on approximately January 15 (see
(Fisher et al., 1987) for more details.) The following code sets up the functional
data object for ξ and then carries out the two integrations required for the two sets
of 35 probe values produced by integrating the product of ξ with each of the basis
functions in each of the two systems.

```
dayvec   = seq(0,365,len=101)
xivec    = exp(20*cos(2*pi*(dayvec-197)/365))
xibasis  = create.bspline.basis(c(0,365),13)
xifd     = smooth.basis(dayvec, xivec, xibasis)$fd
tempLmat = inprod(tempbasis, xifd)
precLmat = inprod(precbasis, xifd)
```

The random behavior of the estimator of whatever we choose to estimate is ul-
timately tied to the random behavior of the data vector $\mathbf{y}$. Let us indicate the order
n variance-covariance matrix of $\mathbf{y}$ as $\text{Var}(y) = \Sigma_e$. Recall that we are operating in
this chapter with the model

$$\mathbf{y} = x(\mathbf{t}) + \varepsilon \,,$$

where $x(\mathbf{t})$ here means the n-vector of values of x at the n argument values t_j. In this model $x(\mathbf{t})$ is regarded as fixed, and as a consequence $\Sigma_e = \mathtt{Var}(\varepsilon)$.

6.5.2 Computing Confidence Limits for Probe Values

We compute confidence limits in this book by a rather classic method: The covariance matrix Σ_ξ of $\xi = \mathbf{A}\mathbf{y}$ is

$$\Sigma_\xi = \mathbf{A}\Sigma_y\mathbf{A}'. \tag{6.3}$$

If the residuals from a smooth of the data have a variance-covariance matrix Σ_e, then we see from $\hat{\mathbf{c}} = \mathtt{y2cMap}\ \mathbf{y}$ that the coefficients will have a variance-covariance matrix

$$\Sigma_c = \mathtt{y2cMap}\ \Sigma_e\ \mathtt{y2cMap}'$$

We use the *conditional* variance of the residuals in this equation because we are only interested in the uncertainty in our estimate of $\mathbf{c}$ that comes from *unexplained* variation in $\mathbf{y}$ after we have explained what we can with our smoothing process. This in turn estimates the random variability in our estimate of the smooth.

We apply (6.3) a second time to get the variance-covariance matrix Σ_ξ for a functional probe by

$$\Sigma_\xi = \mathtt{c2rMap}\ \Sigma_c\ \mathtt{c2rMap}' = \mathtt{c2rMap}\ \mathtt{y2cMap}\ \Sigma_e\ \mathtt{y2cMap}'\ \mathtt{c2rMap}'. \tag{6.4}$$

6.5.3 Confidence Limits for Prince Rupert's Log Precipitation

We can now plot the smooth of the precipitation data for Prince Rupert, British Columbia, Canada's rainiest weather station. The log precipitation data are stored in 365 by 35 matrix `logprecav`, and Prince Rupert is the 29th weather station in our database. We first smooth the data:

```
lambda      = 1e6;
fdParobj    = fdPar(daybasis, harmaccelLfd, lambda)
logprecList= smooth.basis(day.5, logprecav, fdParobj)
logprec.fd = logprecList$fd
fdnames = list("Day (July 1 to June 30)",
          "Weather Station" = CanadianWeather$place,
          "Log 10 Precipitation (mm)")
logprec.fd$fdnames = fdnames
```

Next we estimate Σ_e, which we assume is diagonal. Consequently, we need only estimate the variance of the residuals across weather stations for each day. We do this by smoothing the log of the mean square residuals and then exponentiating the result:

```
logprecmat   = eval.fd(day.5, logprec.fd)
logprecres   = logprecav - logprecmat
logprecvar   = apply(logprecres^2, 1, sum)/(35-1)
lambda       = 1e8
resfdParobj  = fdPar(daybasis, harmaccelLfd, lambda)
logvar.fit   = smooth.basis(day.5, log(logprecvar),
                            resfdParobj)
logvar.fd    = logvar.fit$fd
varvec       = exp(eval.fd(daytime, logvar.fd))
SigmaE       = diag(as.vector(varvec))
```

Next we get y2cMap from the output of smooth.basis, and compute c2rMap by evaluating the smoothing basis at the sampling points. We then compute the variance-covariance matrix for curve values, and finish by plotting the log precipitation curve for Prince Rupert along with this curve plus and minus two standard errors. The result is Figure 6.6.

```
y2cMap = logprecList$y2cMap
c2rMap = eval.basis(day.5, daybasis)
Sigmayhat = c2rMap %*% y2cMap %*% SigmaE %*%
            t(y2cMap) %*% t(c2rMap)
logprec.stderr = sqrt(diag(Sigmayhat))
logprec29 = eval.fd(day.5, logprec.fd[29])
plot(logprec.fd[29], lwd=2, ylim=c(0.2, 1.3))
lines(day.5, logprec29 + 2*logprec.stderr,
      lty=2, lwd=2)
lines(day.5, logprec29 - 2*logprec.stderr,
      lty=2, lwd=2)
points(day.5, logprecav[,29]))
```

6.6 Some Things to Try

1. The 35 Canadian weather stations are divided into four climate zones. These are given in the character vector CanadianWeather$region that is available in the fda package. After computing and plotting the variance-covariance functional data object for the temperature data, compare this with the same analysis applied only to the stations within each region to see if the variability varies between regions. In Chapter 10 we will examine how the mean temperature curves changes from one region to another.
2. What does the covariance bivariate functional data object look like describing the covariation between temperature and log precipitation?

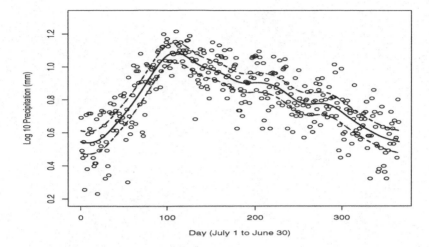

Fig. 6.6 The solid curve is the smoothed base 10 logarithm of the precipitation at Prince Rupert, British Columbia. The dashed lines indicate 95% pointwise confidence limits for the smooth curve based on the data shown as circles.

3. Examine the phase-plane diagram for each of the temperature curves.
4. Compute the standard deviation function for the precipitation data and for the log precipitation data. For each case, plot values of the standard deviation function against values of the mean function. Do you see a general linear trend for the precipitation data and less of that trend for the log precipitation data?
5. Examine the residuals for the growth data from their monotone smooths. Do they appear to be normally distributed or do they exhibit long tails? Do the error variances seem to vary substantially from child to child? Are there any outliers, perhaps due to a failure of the smoothing algorithm, or problems with the measurement process? How does error variance depend on age?
6. Explore the residuals for correlation structure. How would one do this when the data are not equally distributed? One possibility is to treat them as spatial data, and use methods developed in that domain to answer these questions.

Chapter 7
Exploring Variation: Functional Principal and Canonical Components Analysis

Now we look at how observations vary from one replication or sampled value to the next. There is, of course, also variation *within* observations, but we focused on that type of variation when considering data smoothing in Chapter 5.

Principal components analysis, or PCA, is often the first method that we turn to after descriptive statistics and plots. We want to see what primary modes of variation are in the data, and how many of them seem to be substantial. As in multivariate statistics, *eigenvalues* of the bivariate *variance-covariance function* $v(s,t)$ are indicators of the importance of these principal components, and plotting eigenvalues is a method for determining how many principal components are required to produce a reasonable summary of the data.

In functional PCA, there is an *eigenfunction* associated with each eigenvalue, rather than an eigenvector. These eigenfunctions describe major variational components. Applying a rotation to them often results in a more interpretable picture of the dominant modes of variation in the functional data, without changing the total amount of common variation.

We take some time over PCA partly because this may be the most common functional data analysis and because the tasks that we face in PCA and our approaches to them will also be found in more model-oriented tools such as functional regression analysis. For example, we will see that each eigenfunction can be constrained to be smooth by the use of roughness penalties, just as in the data smoothing process. Should we use rough functions to capture every last bit of interesting variation in the data and then force the eigenfunctions to be smooth, or should we carefully smooth the data first before doing PCA?

A companion problem is the analysis of the *covariation* between two different functional variables based on samples taken from the same set of cases or individuals. For example, what types of variation over weather stations do temperature and log precipitation share? How do knee and hip angles covary over the gait cycle? Canonical correlation analysis (CCA) is the method of choice here. We will see many similarities between PCA and CCA.

J.O. Ramsay et al., *Functional Data Analysis with R and MATLAB*, Use R,
DOI: 10.1007/978-0-387-98185-7_7,
© Springer Science + Business Media, LLC 2009

7.1 An Overview of Functional PCA

In multivariate statistics, variation is usually summarized by either the covariance matrix or the correlation matrix. Because the variables in a multivariate observation can vary a great deal in location and scale due to relatively arbitrary choices of origin and unit of measurement, and because location/scale variation tends to be uninteresting, multivariate analyses are usually based on the correlation matrix. But when an observation is functional, values $x_i(s)$ and $x_i(t)$ have the same origin and scale. Consequently, either the estimated *covariance function*

$$v(s,t) = (N-1)^{-1} \sum_i [x_i(s) - \bar{x}(s)][x_i(t) - \bar{x}(t)],$$

or the *cross-product function*

$$c(s,t) = N^{-1} \sum_i x_i(s)x_i(t),$$

will tend to be more useful than the *correlation function*

$$r(s,t) = \frac{v(s,t)}{\sqrt{[v(s,s)v(t,t)]}}.$$

Principal components analysis may be defined in many ways, but its motivation is perhaps clearer if we define PCA as the search for a probe ξ, of the kind that we defined in Chapter 6, that reveals the most important type of variation in the data. That is, we ask, "For what weight function ξ would the probe scores

$$\rho_\xi(x_i) = \int \xi(t)x_i(t)dt$$

have the largest possible variation?" In order for the question to make sense, we have to impose a size restriction on ξ, and it is mathematically natural to require that $\int \xi^2(t)dt = 1$.

Of course, the mean curve by definition is a mode of variation that tends to be shared by most curves, and we already know how to estimate this. Consequently, we usually remove the mean first and then probe the functional residuals $x_i - \bar{x}$. Later, when we look at various types of functional regression, we may also want to first remove other known sources of variation that are explainable by multivariate and/or functional covariates.

The probe score variance $\mathrm{Var}[\int \xi(t)(x_i(t) - \bar{x}(t))^2dt]$ associated with a probe weight ξ is the value of

$$\mu = \max_\xi \{ \sum_i \rho_\xi^2(x_i) \} \text{ subject to } \int \xi^2(t)dt = 1. \tag{7.1}$$

In standard terminology, μ and ξ are referred to as the largest *eigenvalue* and *eigenfunction*, respectively, of the estimated variance-covariance function v. An alternative to the slightly intimidating term "eigenfunction" is *harmonic*.

As in multivariate PCA, a nonincreasing sequence of eigenvalues $\mu_1 \geq \mu_2 \geq \ldots \mu_k$ can be constructed stepwise by requiring each new eigenfunction, computed in step ℓ, to be orthogonal to those computed on previous steps,

$$\int \xi_j(t)\xi_\ell(t)dt = 0, \ j = 1,\ldots,\ell-1 \ \text{and} \ \int \xi_\ell^2(t)dt = 1. \tag{7.2}$$

In multivariate settings the entire suite of eigenvalue/eigenvector pairs would be computed by the *eigenanalysis* of the covariance matrix $\mathbf{V}$, solving the matrix eigenequation $\mathbf{V}\xi_j = \mu_j\xi_j$. The approach is essentially the same for functional data; that is, we calculate eigenfunctions ξ_j of the bivariate covariance function $v(s,t)$ as solutions of the functional eigenequation

$$\int v(s,t)\xi_j(t)dt = \mu_j\xi_j(s). \tag{7.3}$$

We see here as well as elsewhere that going from multivariate to functional data analysis is often only a matter of replacing summation over integer indices by integration over continuous indices such as t. Although the computation details are not at all the same, this is thankfully hidden by the notation and dealt with in the fda package.

However, there is an important difference between multivariate and functional PCA caused by the fact that, whereas in multivariate data the number of variables p is usually less than the number of observations N, for functional data the number of observed function values n is usually greater than N. This implies that the maximum number of nonzero eigenvalues in the functional context is $\min\{N-1,K,n\}$, and in most applications will be $N-1$.

Suppose, then, that our software can present us with, say, $N-1$ positive eigenvalue/eigenfunction pairs (μ_j,ξ_j). What do we do next? For each choice of ℓ, $1 \leq \ell \leq N-1$, the ℓ leading eigenfunctions or harmonics define a basis system that can be used to approximate the sample functions x_i. These basis functions are orthogonal to each other and are normalized in the sense that $\int \xi_\ell^2 = 1$. They are therefore referred to as an *orthonormal* basis. They are also the most efficient basis possible of size ℓ in the sense that the total error sum of squares

$$\text{PCASSE} = \sum_i^N \int [x_i(t) - \bar{x}(t) - \mathbf{c}_i'\xi(t)]^2 dt \tag{7.4}$$

is the minimum achievable with only ℓ basis functions. Of course, other ℓ-dimensional systems certainly exist that will do as well, and we will consider some shortly, but none will do better. In the physical sciences, these optimal basis functions ξ_j are often referred to as *empirical orthogonal functions*.

It turns out that there is a simple relationship between the optimal total squared error and the eigenvalues that are discarded, namely that

$$\text{PCASSE} = \sum_{j=\ell+1}^{N-1} \mu_j.$$

It is usual, therefore, to base a decision on the number ℓ of harmonics to use on a visual inspection of a plot of the eigenvalues μ_j against their indices j, a display that is often referred to in the social science literature as a *scree plot*. Although there are a number of proposals for automatic data-based rules for deciding the value of ℓ, many nonstatistical considerations can also affect this choice.

The coefficient vectors $\mathbf{c}_i, i = 1, \ldots, N$ contain the coefficients c_{ij} that define the optimal fit to each function x_i, and are referred to as *principal component scores*. They are given by the following:

$$c_{ij} = \rho_{\xi_j}(x_i - \bar{x}) = \int \xi_j(t)[x_i(t) - \bar{x}(t)]\mathrm{d}t. \tag{7.5}$$

As we will show below, they can be quite helpful in interpreting the nature of the variation identified by the PCA. It is also common practice to treat these scores as "data" to be subjected to a more conventional multivariate analysis.

We suggested that the eigenfunction basis was optimal but not unique. In fact, for any nonsingular square matrix $\mathbf{L}$ of order ℓ, the system $\phi = \mathbf{T}\xi$ is also optimal and spans exactly the same functional subspace as that spanned by the eigenfunctions. Moreover, if $\mathbf{T}' = \mathbf{T}^{-1}$, such matrices being often referred to as *rotation matrices*, the new system ϕ is also orthonormal. There is, in short, no mystical significance to the eigenfunctions that PCA generates, a simple fact that is often overlooked in textbooks on multivariate statistics. Well, okay, perhaps $\ell = 1$ is an exception. In fact, it tends to happen that only the leading eigenfunction has an obvious meaningful interpretation in terms of processes known to generate the data.

But for $\ell > 1$, there is nothing to prevent us from searching among the infinite number of alternative systems $\phi = \mathbf{T}\xi$ to find one where all of the orthonormal basis functions ϕ_j are seen to have some substantive interpretation. In the social sciences, where this practice is routine, a number of criteria for optimizing the chances of interpretability have been devised for choosing a rotation matrix $\mathbf{T}$, and we will demonstrate the usefulness of the popular *VARIMAX* criterion in our examples.

Readers are referred at this point to standard texts on multivariate data analysis or to the more specialized treatment in Jolliffe (2002) for further information on principal components analysis. Most of the material in these sources applies to this functional context.

7.2 PCA with Function pca.fd

Principal component analysis is implemented in the functions *pca.fd* and pca_fd in R and Matlab, respectively. The call in R is

```
pca.fd(fdobj, nharm = 2, harmfdPar=fdPar(fdobj),
```

```
                     centerfns = TRUE)
```

The first argument is a functional data object containing the functional data to be analyzed, and the second specifies the number ℓ of principal components to be retained. The third argument is a functional parameter object that provides the information necessary to smooth the eigenfunctions if necessary; we will postpone this topic to Section 7.3. Finally, although most principal components analyses are applied to data with the mean function subtracted from each function, the final argument permits this to be suppressed.

Function `pca.fd` in R returns an object with the class name `pca.fd`, so that it is effectively a constructor function. Here are the named components for this class:

harmonics   A functional data object for the ℓ harmonics or eigenfunctions ξ_j.
values   The complete set of eigenvalues μ_j.
scores   The matrix of scores c_{ij} on the principal components or harmonics.
varprop   A vector giving the proportion $\mu_j / \sum \mu_j$ of variance explained by each
    eigenfunction.
meanfd   A functional data object giving the mean function $\bar{x}$.

7.2.1 PCA of the Log Precipitation Data

Here is the command to do a PCA using only two principal components for the log precipitation data and to display the eigenvalues.

```
logprec.pcalist = pca.fd(logprecfd, 2)
print(logprec.pcalist$values)
```

We observe that these two harmonics account for 96% of the variation around the mean log precipitation curve; the first four eigenvalues are 39.5, 3.9, 1.0 and 0.4, respectively.

The two principal components are plotted by the command

```
plot.pca.fd(logprec.pcalist)
```

Figure 7.1 shows the two principal component functions by displaying the mean curve along +'s and −'s indicating the consequences of adding and subtracting a small amount of each principal component. We do this because a principal component represents *variation* around the mean, and therefore is naturally plotted as such. We see that the first harmonic, accounting for 88% of the variation, represents a relative constant vertical shift in the mean, and that the second shows essentially a contrast between winter and summer precipitation levels.

It is in fact usual for unrotated functional principal components to display the same sequence of variation no matter what is being analyzed. The first will be a constant shift, the second a linear contrast between the first and second half with a single crossing of zero, the third a quadratic pattern, and so on. That is, we tend to see the sequence of orthogonal polynomials. However, for periodic data, where only periodic harmonics are possible, the linear contrast is suppressed.

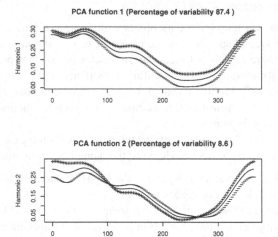

Fig. 7.1 The two principal component functions or harmonics are shown as perturbations of the mean, which is the solid line. The +'s show what happens when a small amount of a principal component is added to the mean, and the −'s show the effect of subtracting the component.

The fact that unrotated functional principal components are so predictable emphasizes the need for looking for a rotation of them that can reveal more meaningful components of variation. The VARIMAX rotation algorithm is often used for this purpose. The following command applies this rotation and then plots the result:

```
logprec.rotpcalist = varmx.pca.fd(logprec.pcalist)
plot.pca.fd(logprec.rotpcalist)
```

The results are plotted in Figure 7.2. The first component portrays variation that is strongest in midwinter and the second captures primarily summer variation.

It can be profitable to plot the principal component scores for pairs of harmonics to see how curves cluster and otherwise distribute themselves within the K-dimensional subspace spanned by the eigenfunctions. Figure 7.3 reveals some fascinating structure. Most of the stations are contained within two clusters: the upper right with the Atlantic and central Canada stations and the lower left with the prairie and mid-Arctic stations. The outliers are the three west coast stations and Resolute in the high Arctic. Often, functional data analyses will turn into a multivariate data analysis at this point by using the component scores as "data matrices" in more conventional analyses.

It may be revealing to apply PCA to some order of derivative rather than to the curves themselves, because underlying processes may reveal their effects at the change level rather than at the level of what we measure. This is certainly true of growth curve data, where hormonal processes and other growth activators change

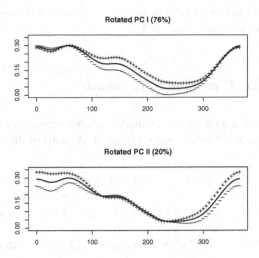

Fig. 7.2 The two rotated principal component functions are shown as perturbations of the mean, which is the solid line. The top panel contains the strongest component, with variation primarily in the midwinter. The bottom panel shows primarily summer variation.

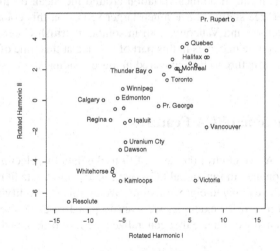

Fig. 7.3 The scores for the two rotated principal component functions are shown as circles. Selected stations are labeled in order to identify the two central clusters and the outlying stations.

the rate of change of height and can be especially evident at the level of the acceleration curves that we plotted in Section 1.1.

7.2.2 PCA of Log Precipitation Residuals

We can now return to exploring the residuals from the smooths of the log precipitation curves in Chapter 5. First, we set up function versions of the residuals and plot them:

```
logprecres.fd = smooth.basis(day.5, logprecres,
                             fdParobj)$fd
plot(logprecres.fd, lwd=2, col=1, lty=1, cex=1.2,
     xlim=c(0,365), ylim=c(-0.07, 0.07),
     xlab="Day", ylab="Residual (log 10 mm)")
```

These are shown in Figure 7.4. There we see that, while most of these residual functions show fairly chaotic variation, three stations have large oscillations in summer and autumn. The result of estimating a single principal component is shown in Figure 7.5, where we see the mean residual along with the effect of adding and subtracting this first component. The mean residual itself shows the oscillation that we have noted. The principal component accounts for about 49% of the residual variance about this mean. It defines variation around the mean oscillation located in these months. Three stations have much larger scores on this component: They are Kamloops, Victoria and Vancouver, all in southern British Columbia. It seems that rainfall events come in cycles in this part of Canada at this time of the year, and there is interesting structure to be uncovered in these residuals.

7.3 More Functional PCA Features

In multivariate PCA, we control the level of fit to the data by selecting the number of principal components. In functional PCA, we can also modulate fit by controlling the roughness of the estimated eigenfunctions. We do this by modifying the definition of orthogonality. If, for example, we want to penalize excessive curvature in principal components, we can use this generalized form of orthogonality:

$$\int \xi_j(t)\xi_k(t)dt + \lambda \int D^2\xi_j(t)D^2\xi_k(t)dt = 0, \tag{7.6}$$

where λ controls the relative emphasis on orthogonality of second derivatives in much the same way as it does in roughness–controlled smoothing. This gives us a powerful new form of leverage in defining a decomposition of variation.

Roughness-penalized PCA also relates to a fundamental aspect of variation in function spaces. Functions can be large in two distinct ways: first and most obvi-

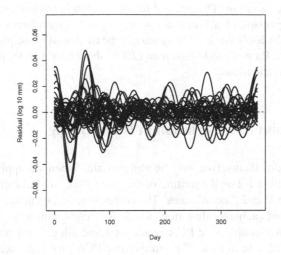

Fig. 7.4 The smoothed residual functions for the log precipitation data.

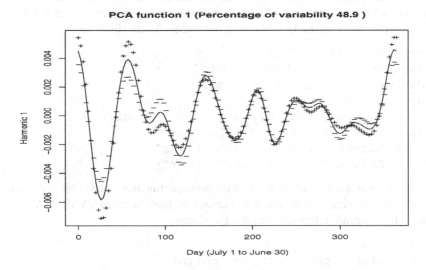

Fig. 7.5 The first principal component for the log precipitation residual functions, shown by adding (+) and subtracting (-) the component from the mean function (solid line).

ously in terms of their amplitude, and second in terms of their complexity or amount of high-frequency variation. This second feature is closely related to how rapidly a Fourier series expansion of a function converges, and is therefore simply another aspect of how PCA itself works. This second type of size of principal components is what λ controls. Ramsay and Silverman (2005) show how λ in PCA can be data-defined via cross-validation.

7.4 PCA of Joint X-Y Variation in Handwriting

Of course, functions themselves may be multivariate. When we apply PCA to the data shown in Section 1.2 on the writing of the script "fda," we have to do a simultaneous PCA of the X and Y coordinates. The corresponding eigenfunctions will also be multivariate, but each eigenfunction is still associated with a single eigenvalue μ_j. This means that multivariate PCA is not the same thing as separate PCA's applied to each coordinate in turn. The multivariate PCA problem, therefore, blends together the aspects of multivariate and functional data analyses.

At the level of code, however, multivariate PCA is achieved seamlessly by function `pca.fd`. These R commands define a small but sufficient number of basis functions for representing the "fda" handwriting data as a bivariate functional data object, smooth the data, and install appropriate labels for the dimensions.

```
fdarange = c(0, 2300)
fdabasis = create.bspline.basis(fdarange, 105, 6)
fdatime  = seq(0, 2300, len=1401)
fdafd =
   smooth.basis(fdatime, handwrit, fdabasis)$fd
fdafd$fdnames[[1]] = "Milliseconds"
fdafd$fdnames[[2]] = "Replications"
fdafd$fdnames[[3]] = list("X", "Y")
```

These R commands carry out the PCA of the bivariate functional data object `fdafd` using three harmonics, plot the unrotated eigenfunctions, perform a VARIMAX rotation of these eigenfunctions, and replot the results.

```
nharm = 3
fdapcaList = pca.fd(fdafd, nharm)
plot.pca.fd(fdapcaList)
fdarotpcaList = varmx.pca.fd(fdapcaList)
plot.pca.fd(fdarotpcaList)
```

How did we settle on three for the number of harmonics? We have found that the logarithm of eigenvalues tend to decrease linearly after an initial few that are large. The following commands plot the log eigenvalues up to $j = 12$ with the least-squares linear trend in the eigenvalue with indices 4 to 12.

```
fdaeig = fdapcaList$values
```

```
neig    = 12
x       = matrix(1,neig-nharm,2)
x[,2]   = (nharm+1):neig
y       = log10(fdaeig[(nharm+1):neig])
c       = lsfit(x,y,int=FALSE)$coef
par(mfrow=c(1,1),cex=1.2)
plot(1:neig, log10(fdaeig[1:neig]), "b",
     xlab="Eigenvalue Number",
     ylab="Log10 Eigenvalue")
lines(1:neig, c[1]+ c[2]*(1:neig), lty=2)
```

The result is Figure 7.6. The first three log eigenvalues seem well above the linear trend in the next nine, suggesting that the leading three harmonics are important. Together they account for 62% of the variation in the scripts.

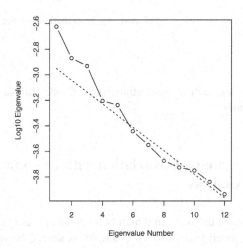

Fig. 7.6 The logarithms (base 10) of the first 12 eigenvalues in the principal components analysis of the "fda" handwriting data. The dashed line indicates the linear trend in the last nine in the sequence.

Figure 7.7 plots two of the VARIMAX–rotated eigenfunctions as perturbations of the mean script. The rotated harmonic on the left mostly captures variation in the lower loop of "f", and the harmonic on the right displays primarily variation in its upper loop. This suggests that variabilities in these two loops are independent of each other.

We can also analyze situations where there are both functional and multivariate data available, such as handwritings from many subjects along with measurements

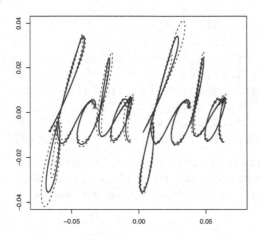

Fig. 7.7 Two of the rotated harmonics are plotted as a perturbations of the mean "fda" script, shown as a heavy solid line.

of subject characteristics such as age, ethnicity, etc. See Ramsay and Silverman (2005) for further details.

7.5 Exploring Functional Covariation with Canonical Correlation Analysis

We often want to examine the ways in which two sets of curves $(x_i, y_i), i = 1, \ldots, N$, share variation. How much variation, for example, is shared between temperature and log precipitation over the 35 Canadian weather stations? This question is related to the issue of how well one can predict one from another, which we will take up in the next chapter. Here, we consider a symmetric view on the matter that does not privilege either variable. We offer here only a quick summary of the mathematical aspects of canonical correlation analysis, and refer the reader to Ramsay and Silverman (2005) for a more detailed account.

To keep the notation tidy, we will assume that the two sets of variables have been *centered*, that is, x_i and y_i have been replaced by the residuals $x_i - \bar{x}$ and $y_i - \bar{y}$, respectively, if this was considered appropriate. That is, we assume that $\bar{x} = \bar{y} = 0$. As before, we define modes of variation for the x_i's and the y_i's in terms of the pair of *probe weight functions* ξ and η that define the integrals

$$\rho_{\xi i} = \int \xi(t) x_i(t) \mathrm{d}t \quad \text{and} \quad \rho_{\eta i} = \int \eta(t) y_i(t) \mathrm{d}t, \tag{7.7}$$

respectively. The N pairs of *probe scores* $(\rho_{\xi i}, \rho_{\eta i})$ defined in this way represent shared variation if they correlate strongly with one another.

The *canonical correlation criterion* is the squared correlation

$$R^2(\xi, \eta) = \frac{[\Sigma_i \rho_{\xi i} \rho_{\eta i}]^2}{[\Sigma_i \rho_{\xi i}^2][\Sigma_i \rho_{\eta i}^2]} = \frac{[\Sigma_i (\int \xi(t) x_i(t) dt)(\int \eta(t) y_i(t) dt)]^2}{[\Sigma_i (\int \xi(t) x_i(t) dt)^2][\Sigma_i (\int \eta(t) y_i(t) dt)^2]}. \tag{7.8}$$

As in PCA, the probe weights ξ and η are then specified by finding that weight pair that optimizes the criterion $R^2(\xi, \eta)$. But, again as in PCA, we can compute a nonincreasing series of squared canonical correlations $R_1^2, R_2^2, \ldots, R_k^2$ by constraining successive canonical probe values to be orthogonal. The length k of the sequence is the smallest of the sample size N, the number of basis functions for either functional variable, or the number of basis functions used for ξ and η.

That we are now optimizing with respect to two probes at the same time makes canonical correlation analysis an exceedingly *greedy* procedure, where this term borrowed from data mining implies that CCA can capitalize on the tiniest variation in either set of functions in maximizing this ratio to the extent that, unless we exert some control over the process, it can be hard to see anything of interest in the result. It is in practice essential to enforce strong smoothness on the two weight functions ξ and η to limit this greediness. This can be done by either selecting a low-dimensional basis for each or by using an explicit roughness penalty in much the same manner as is possible for functional PCA.

Let us see how this plays out in the exploration of covariation between daily temperature and log precipitation, being careful to avoid the greediness pitfall by placing very heavy penalties on roughness of the canonical weight functions as measured by the size of their second derivatives. Here are the commands in R that function cca.fd to do the job:

```
ccafdPar  = fdPar(daybasis, 2, 5e6)
ncon      = 3
ccalist   = cca.fd(temp.fd, logprec.fd, ncon,
                    ccafdPar, ccafdPar)
```

The third argument of cca.fd specifies the number of canonical weight/variable pairs that we want to examine, which, in this case, is the complete sequence. The final two arguments specify the bases for the expansion of ξ and η, respectively, as well as their roughness penalties.

The canonical weight functional data objects and the corresponding three squared canonical correlations are extracted from the list object ccalist produced by function cca.fd as follows:

```
ccawt.temp    = ccalist$ccawtfd1
ccawt.logprec = ccalist$ccawtfd2
corrs         = ccalist$ccacorr
```

The squared correlations are 0.92, 0.62 and 0.35; so that there is a dominant pair of modes of variation that correlates at a high level, and then two subsequent pairs with modest but perhaps interesting correlations.

Consider first the type of variation associated with the first canonical correlation. Figure 7.8 displays the corresponding two canonical weight functions. The temperature canonical weight function ξ_1 resembles a sinusoid with period 365/2 and having zeros in July, October, January and April. But the log precipitation counterpart η_1 is close to a sinusoid with period 365 and zeros in July and January.

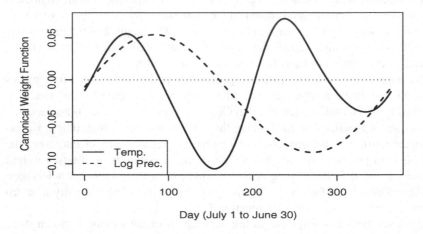

Fig. 7.8 The first pair of canonical weight functions or probes (ξ, η) correlating temperature and log precipitation for the Canadian weather data.

Regarding each weight function as contrasting corresponding variable values, the temperature curve seems primarily to contrast spring and autumn temperatures with winter temperatures; while the corresponding log precipitation contrast is between rainfall in the spring and autumn. A station will score high on both canonical variables if it is cool in winter relative to its temperatures in spring and autumn, and at the same time has more precipitation in the spring than in the fall.

The scores of each weather station on each set of canonical variables are extracted by

```
ccascr.temp    = ccalist$ccavar1
ccascr.logprec = ccalist$ccavar2
```

Figure7.9 plots the scores for the first log precipitation canonical variable scores against their temperature counterparts for selected weather stations. We see a near-perfect ordering with respect to latitude, although favoring eastern stations over western stations at the same latitudes so that Vancouver and Victoria wind up at the bottom left. Certainly Resolute's temperatures are cold in winter, and what precipitation it gets comes more in the spring than at another time, so that it earns it's place in the upper right of the plot. The marine weather stations, Prince Rupert and

St. John's, on the other hand, are actually relatively warm in the winter and get more precipitation in the fall than in the winter, and therefore anchor the lower left of the plot. Note, though, that the linear order in Figure7.9 misses Kamloops by a noticeable amount. The position of this interior British Columbia city deep in a valley, where relatively little rain or snow falls at any time of the year, causes it to be anomalous in many types of analysis.

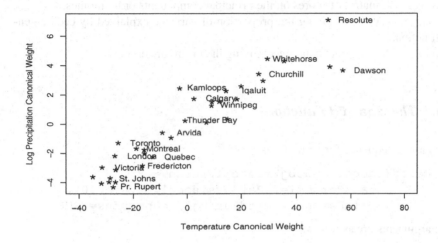

Fig. 7.9 The scores for the first pair of canonical variables plotted against each other, with labels for selected weather stations.

7.6 Details for the pca.fd and cca.fd Functions

7.6.1 The pca.fd Function

We give here the arguments of the constructor function pca.fd that carries out a functional principal components analysis and constructs an object of the pca.fd class. The complete calling sequence is

```
pca.fd(fdobj, nharm = 2, harmfdPar=fdPar(fdobj),
       centerfns = TRUE)
```

The arguments are as follows:

fdobj   A functional data object.
nharm   The number of harmonics or principal components to compute.

harmfdPar   A functional parameter object that defines the harmonic or principal component functions to be estimated.

centerfns   A logical value: if TRUE, subtract the mean function from each function before computing principal components.

Function pca.fd returns an argument of the pca.fd class, which is a named list with the following components:

harmonics   A functional data object for the harmonics or eigenfunctions.

values   The complete set of eigenvalues.

scores   A matrix of scores on the principal components or harmonics.

varprop   A vector giving the proportion of variance explained by each eigenfunction.

meanfd   A functional data object giving the mean function.

7.6.2 The cca.fd Function

The calling sequence for cca.fd is

```
cca.fd(fdobj1, fdobj2=fdobj1, ncan = 2,
       ccafdParobj1=fdPar(basisobj1, 2, 1e-10),
       ccafdParobj2=ccafdParobj1, centerfns=TRUE)
```

The arguments are as follows:

fdobj1   A functional data object.

fdobj2   A functional data object. By default this is fdobj1, in which case the first argument must be a bivariate functional data object.

ncan   The number of canonical variables and weight functions to be computed. The default is 2.

ccafdParobj1   A functional parameter object defining the first set of canonical weight functions. The object may contain specifications for a roughness penalty. The default is defined using the same basis as that used for fdobj1 with a slight penalty on its second derivative.

ccafdParobj2   A functional parameter object defining the second set of canonical weight functions. The object may contain specifications for a roughness penalty. The default is ccafdParobj1.

centerfns   If TRUE, the functions are centered prior to analysis. This is the default.

7.7 Some Things to Try

1. **Medfly Data**: The medfly data have been a popular dataset for functional data analysis and are included in the fda package. The medfly data consist of records

of the number of eggs laid by 50 fruit flies on each of 31 days, along with each individual's total lifespan.

   a. Smooth the data for the number of eggs, choosing the smoothing parameter by generalized cross-validation (GCV). Plot the smooths.
   b. Conduct a principal components analysis using these smooths. Are the components interpretable? How many do you need to retain to recover 90% of the variation. If you believe that smoothing the PCA will help, do so.
   c. Try a linear regression of lifespan on the principal component scores from your analysis. What is the R^2 for this model? Does lm find that the model is significant? Reconstruct and plot the coefficient function for this model along with confidence intervals. How does it compare to the model obtained through functional linear regression?

2. Apply principal components analysis to the functional data object Wfd returned by the monotone smoothing function smooth.monotone applied to the growth data. These functions are the logs of the first derivatives of the growth curves. What is the impact of the variation in the age of the purbertal growth spurt on these components?

7.8 More to Read

Functional principal components analysis predates the emergence of functional data analysis, especially in fields in engineering and sciences that work with functional data routinely, such as climatology. Principal components are often referred to in these fields as *empirical basis functions*, a phrase that is exactly the right thing since functional principal components are both orthogonal and can also serve well as a customized low-dimensional basis system for representing the actual functions.

There are many currently active and unexplored areas of research into functional PCA. James et al. (2000) consider situations where curves are observed in fragments, so that the interval of observation varies from record to record. James and Sugar (2003) look at the same data situation in the context of cluster analysis, another multivariate exploratory tool that is now associated with a large functional literature. Readers with a background in psychometrics will wonder about a functional version of factor analysis, whether exploratory or confirmatory; and functional versions of structural equation models are well down the road, but no doubt perfectly feasible.

Chapter 8
Registration: Aligning Features for Samples of Curves

This chapter presents two methods for separating phase variation from amplitude variation in functional data: landmark and continuous registration. We mentioned this problem in Section 1.1.1. We saw in the height acceleration curves in Figure 1.2 that the age of the pubertal growth spurt varies from girl to girl; this is *phase variation*. In addition, the intensity of the pubertal growth spurt also varies; this is *amplitude variation*. Landmark registration aligns features that are visible in all curves by estimating a strictly increasing nonlinear transformation of time that takes all the times of a given feature into a common value. Continuous registration uses the entire curve rather than specified features and can provide a more complete curve alignment. The chapter also describes a decomposition technique that permits the expression of the amount of phase variation in a sample of functional variation as a proportion of total variation.

8.1 Amplitude and Phase Variation

Figure 1.2 presented the problem that curve registration is designed to solve. This figure is reproduced in the top panel of Figure 8.1 along with a solution in the bottom panel. In both panels, the dashed line indicates the mean of these ten growth acceleration curves. In the top panel, this mean curve is unlike any of the individual curves in that the duration of the mean pubertal growth is longer than it should be and the drop in acceleration is not nearly as steep as even the shallowest of the individual curves. These aberrations are due to the ten girls not being in the same phase of growth at around 10 to 12 years of age. We see from the figure that peak growth acceleration occurs around age 10.5 for many girls, but this occurred before age 8 for one girl and after age 13 for another. Similarly, the maximum pubertal growth rate occurs where the acceleration drops to zero following the maximum pubertal acceleration. This occurs before age 10 for two girls and around age 14 for another, averaging around 11.7 years of age. If we average the growth accelerations at that age, one girl has not yet begun her pubertal growth spurt, three others are

J.O. Ramsay et al., *Functional Data Analysis with R and MATLAB*, Use R,
DOI: 10.1007/978-0-387-98185-7_8,
© Springer Science + Business Media, LLC 2009

at or just past their peak acceleration, and the rest are beyond their peak pubertal growth rate with negative acceleration. This analysis should make it fairly easy to understand why the average of these acceleration curves displays an image that is very different from any of the individual curves.

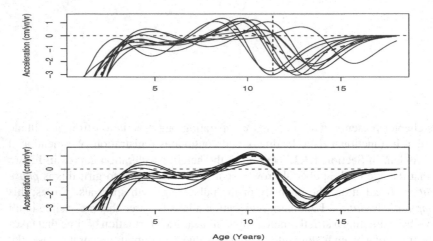

Fig. 8.1 The top panel reproduces the second derivatives of the growth curves shown in Figure 1.2. The landmark–registered curves corresponding to these are shown in the bottom panel, where the single landmark was the crossing of zero in the middle of the mean pubertal growth spurt. The dashed line in each panel indicates the mean curve for the curves in that panel.

The bottom panel in Figure 8.1 uses landmark registration to align these curves so the post–spurt accelerations for all girls cross zero at the same time. Then when we average the curves, we get a much more realistic representation of the typical pubertal growth spurt, at least among the girls in this study.

Functions can vary in both phase and amplitude, as illustrated schematically in Figure 8.2. *Phase variation* is illustrated in the top panel as a variation in the location of curve features along the horizontal axis, as opposed to *amplitude variation*, shown in the bottom panel as the size of these curves. The mean curve in the top panel, shown as a dashed curve, does not resemble any curve; it has less amplitude variation, but its horizontal extent is greater than that of any single curve. The mean has, effectively, borrowed from amplitude to accommodate phase. Moreover, if we carry out a functional principal components analysis of the curves in each panel, we find in the top panel that the first three principal components account for 55%, 39% and 5%, of the variation. On the other hand, the same analysis of the amplitude-varying curves requires a single principal component to account for 100% of the variation. Like the mean and principal components, most statistical methods when translated into the functional domain are designed to model purely amplitude variation.

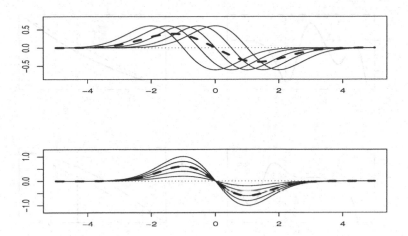

Fig. 8.2 The top panel shows five curves varying only in phase. The bottom panel shows five curves varying only in amplitude. The dashed line in each panel indicates the mean of the five curves. This curve in the bottom panel is superimposed exactly on the central curve.

There is physiological *growth time* that unrolls at different rates from child to child relative to *clock time*. In terms of growth time, all girls experience puberty at the same age, with the peak growth rate (zero acceleration) occurring at about 11.7 years of age for the Berkeley sample. If we want a reasonable sense of amplitude variation, we must consider it with this *growth time* frame of reference. Growth time itself is an elastic medium that can vary randomly from girl to girl when viewed relative to clock time, and functional variation has the potential to be *bivariate*, with variation in both the *range* and *domain* of a function.

8.2 Time-Warping Functions and Registration

We can remove phase variation from the growth data if we can estimate a *time-warping function* $h_i(t)$ that transforms growth time t to clock time for child i. For example, we can require that $h_i(11.7) = t_i$ for all girls, where 11.7 years is the average time at which the Berkeley girls reached their midpubertal spurt (PGS) and t_i is the clock age at which the ith girl reached this event. If, at any time t, $h_i(t) < t$, we may say that the girl is growing faster than average at that clock time but slower than average if $h_i(t) > t$. This is illustrated in Figure 8.3, where the growth acceleration curves for the earliest and latest of the first ten girls are shown in the left panels and their corresponding time-warping functions in the right panels.

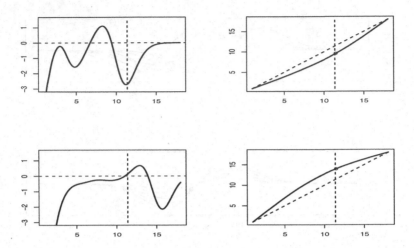

Fig. 8.3 The top panels show the growth acceleration curve on the left and the corresponding time-warping function $h(t)$ on the right for the girl among the first ten in the Berkeley growth study with the earliest pubertal growth spurt. The corresponding plots for the girl with the latest growth spurt are in the bottom two panels. The middle of the growth spurt is shown as the vertical dashed line in all panels.

Time-warping functions must, of course, be strictly increasing; we cannot allow time to go backwards in either frame of reference. Time-warping functions must also be smooth in the sense of being differentiable up to at least what applies to the curves being registered. If the curves are observed over a common interval $[0,T]$, the time-warping functions must often satisfy the constraints $h(0) = 0$ and $h(T) = T$, but it may be that varying intervals $[0,T_i]$ may each be transformed to a common interval $[0,T]$. In the special case of periodic curves, such as average temperature and precipitation profiles, we may also allow a constant shift $h_i(t) = t_i + \delta_i$.

The *registered* height functions are $x_i^*(t) = x_i[h_i^{-1}(t)]$, where the *aligning function* $h^{-1}(t)$ satisfies the following equation:

$$h^{-1}[h(t)] = t. \tag{8.1}$$

This is the *functional inverse* of $h(t)$.

For example, since at time $h_i(t_0)$ girl i is in the middle of her pubertal growth spurt, and since in her registered time $h_i^{-1}[h_i(t_0)] = t_0$, she and all the other children will experience puberty at time t_0 in terms of registered or "growth" time. In particular, if $h_i(t_0) < t_0$ for a girl i reaching puberty early, then aligning function $h_i^{-1}(t)$

effectively slows down or stretches out her clock time so as to conform with growth time.

8.3 Landmark Registration with Function landmarkreg

The simplest curve alignment procedure is *landmark* registration. A landmark is a feature with a location that is clearly identifiable in all curves. Landmarks may be the locations of minima, maxima or crossings of zero, and we see three such landmarks in each curve Figure 8.2. We align the curves by transforming t for each curve so that landmark locations are the same for all curves.

For the bottom panel in Figure 1.2, we used a single landmark t_i being the age for girl i at which her acceleration curve crossed 0 with a negative slope during the pubertal growth spurt. Also, let us define t_0 as a time specified for the middle of the *average* pubertal growth spurt, such as 11.7 years of age for the Berkeley growth study girls. Then we specify time-warping functions h_i by fitting a smooth function to the three points $(1,1), (t_0, t_i)$, and $(18, 18)$. This function should be as differentiable as the curves themselves, and in this case could be simply the unique parabola passing through the three points $(1,1), (t_0, t_i)$ and $(18, 18)$, which is what is shown in the right panels of Figure 8.3.

The code for smoothing the Berkeley female growth data is found in Section 5.4.2.2. From these smooths, we can compute the unregistered accelerations functions and their mean function by the commands

```
accelfdUN     = deriv.fd(hgtfhatfd, 2)
accelmeanfdUN = mean(accelfdUN)
```

This code allows you to select the age of the center of the pubertal growth spurt for each girl, applying the R function locator() to plots of a functional data object accfd that contains estimated acceleration curves.

```
PGSctr  = rep(0,10)
agefine = seq(1,18,len=101)
par(mfrow=c(1,1), ask=TRUE)
for (icase in 1:54) {
    accveci = predict(accelfdUN[icase], agefine)
    plot(agefine,accveci,"l", ylim=c(-6,4),
         xlab="Year", ylab="Height Accel.",
         main=paste("Case",icase))
    lines(c(1,18),c(0,0),lty=2)
    PGSctr[icase] = locator(1)$x
}
PGSctrmean = mean(PGSctr)
```

We don't need much flexibility in the function fitting the three points defining each warping function, so we define four order three spline basis functions and apply a very light level of smoothing in these commands:

```
wbasisLM = create.bspline.basis(c(1,18), 4, 3,
                             c(1,PGSctrmean,18))
WfdLM     = fd(matrix(0,4,1),wbasisLM)
WfdParLM = fdPar(WfdLM,1,1e-12)
```

The landmark registration using function `landmarkreg` along with the extraction of the registered acceleration functions, warping function and w-functions is achieved by the commands

```
regListLM = landmarkreg(accelfdUN, PGSctr,
                      PGSctrmean, WfdParLM, TRUE)
accelfdLM       = regListLM$regfd
accelmeanfdLM = mean(accelfdLM)
warpfdLM        = regList$warpfd
WfdLM           = regList$Wfd
```

The final logical argument value TRUE requires the warping functions h_i to themselves be strictly monotone functions.

The bottom panel of Figure 8.1 displays the same ten female growth acceleration curves after registering to the middle of the pubertal growth spurt. We see that the curves are now exactly aligned at the mean PGS (pubertal growth spurt) age, but that there is still some misalignment for the maximum and minimum acceleration ages. Our eye is now drawn to the curve for girl seven, whose acceleration minimum is substantially later than the others and who has still not reached zero acceleration by age 18. The long period of near zero acceleration for girl four prior to puberty also stands out as unusual. The mean curve is now much more satisfactory as a summary of the typical shape of growth acceleration curves, and in particular is nicely placed in the middle of the curves for the entire pubertal growth spurt period.

8.4 Continuous Registration with Function `register.fd`

We may need registration methods that use the entire curves rather than their values at specified points. A number of such methods have been developed, and the problem continues to be actively researched. Landmark registration is usually a good first step, but we need a more refined registration process if landmarks are not visible in all curves. For example, many but not all female growth acceleration curves have at least one peak prior to the pubertal growth spurt that might be considered a landmark. Even when landmarks are clear, identifying their timing may involve tedious interactive graphical procedures, and we might prefer a fully automatic method. Finally, as we saw in Figure 8.1, landmark registration using just a few landmarks can still leave aspects of the curves unregistered at other locations.

Here we illustrate the use of function `register.fd` to further improve the acceleration curves that have already been registered using function `landmarkreg`. The idea behind this method is that if an arbitrary sample registered curve $x[h(t)]$ and target curve $x_0(t)$ differ only in terms of amplitude variation, then their values will tend to be proportional to one another across the range of t-values. That is, if we were to plot the values of the registered curve against the target curve, we would see something approaching a straight line tending to pass through the origin, although not necessarily at angle 45 degrees with respect to the axes of the plot. If this is true, then a principal components analysis of the following order two matrix $\mathbf{T}(h)$ of integrated products of these values should reveal essentially one component, and the smallest eigenvalue should be near 0:

$$\mathbf{C}(h) = \begin{bmatrix} \int \{x_0(t)\}^2 \, dt & \int x_0(t)x[h(t)] \, dt \\ \int x_0(t)x[h(t)] \, dt & \int \{x[h(t)]\}^2 \, dt \end{bmatrix}. \tag{8.2}$$

According to this rationale, then, estimating h so as to minimize the smallest eigenvalue of $\mathbf{C}(h)$ should do the trick. This is exactly what `register.fd` does for each curve in the sample.

If the curves are multivariate, such as coordinates of a moving point, then what is minimized is the sum of the smallest eigenvalues across the components of the curve vectors. We recall, too, that curve $x(t)$ may in fact be a derivative of the curve used to smooth the data.

In the following code, we use a more powerful basis than we used in Chapter 5, combined with a roughness penalty, for defining the functions $W(t)$ to estimate strictly monotone functions. Because the continuous registration process requires iterative numerical optimization techniques, we have to supply starting values for the coefficients defining the functions W. We do this by using zeros in defining the initial functional data object `Wfd0CR`.

```
wbasisCR = create.bspline.basis(c(1,18), 15, 5)
Wfd0CR   = fd(matrix(0,15,54),wbasisCR)
WfdParCR = fdPar(Wfd0CR, 1, 1)
regList  = register.fd(mean(accelfdLM),
                       accelfdLM, WfdfdPar)
accelfdCR = regList$regfd
warpfdCR  = regList$warpfd
WfdCR     = regList$Wfd
```

Figure 8.4 shows that the continuously registered height acceleration curves are now aligned over the entire PGS relative to the landmark–registered curves, although we have sacrificed a small amount of alignment of the zero–crossings of these curves. Figure 8.5 shows the impacts of the two types of registrations, and we see that, while both registrations provide average curves with maximum and minimum values much more typical of the individual curves, as well as the width of the PGS, the final continuously registered mean curve does a better job in the mid-spurt period centered on five years of age.

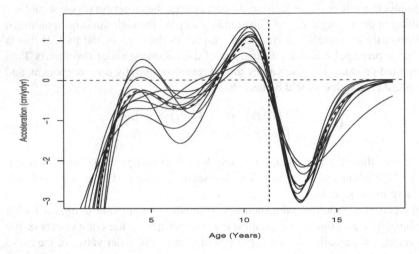

Fig. 8.4 The continuous registration of the landmark–registered height acceleration curves in Figure 8.1. The vertical dashed line indicates the target landmark age used in the landmark registration.

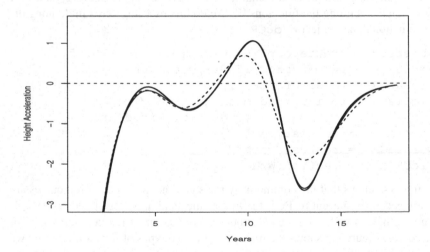

Fig. 8.5 The mean of the continuously registered acceleration curves is shown as a heavy solid line, while that of the landmark-registered curves is a light solid line. The light dashed line is the mean of the unregistered curves

8.5 A Decomposition into Amplitude and Phase Sums of Squares

Kneip and Ramsay (2008) developed a useful way of quantifying the amount of these two types of variation by comparing results for a sample of N functional observations before and after registration. The notation x_i stands for the unregistered version of the ith observation, y_i for its registered counterpart and h_i for associated warping function. The sample means of the unregistered and registered samples are $\bar{x}$ and $\bar{y}$, respectively.

The *total mean square error* is defined as

$$MSE_{total} = N^{-1} \sum_i^N \int [x_i(t) - \bar{x}(t)]^2 \, dt. \tag{8.3}$$

We define the constant C_R as

$$C_R = 1 + \frac{N^{-1} \sum_i^N \int [Dh_i(t) - N^{-1} \sum_i^N Dh_i(t)][y_i^2(t) - N^{-1} \sum_i^N y_i^2(t)] \, dt}{N^{-1} \sum_i^N \int y_i^2(t) \, dt}. \tag{8.4}$$

The structure of C_R indicates that $C_R - 1$ is related to the covariation between the deformation functions Dh_i and the squared registered functions y_i^2. When these two sets of functions are independent, the number of the ratio is 0 and $C_R = 1$.

The measures of amplitude and phase mean square error are, respectively,

$$MSE_{amp} = C_R N^{-1} \sum_i^N \int [y_i(t) - \bar{y}(t)]^2 \, dt$$

$$MSE_{phase} = C_R \int \bar{y}^2(t) \, dt - \int \bar{x}^2(t) \, dt. \tag{8.5}$$

It can be shown that, defined in this way, $MSE_{total} = MSE_{amp} + MSE_{phase}$.

The interpretation of this decomposition is as follows. If we have registered our functions well, then the registered functions y_i will have higher and sharper peaks and valleys, since the main effect of mixing phase variation with amplitude variation is to smear variation over a wider range of t values, as we saw in Figure 1.2 and Figure 8.2. Consequently, the first term in MSE_{phase} will exceed the second and is a measure of how much phase variation has been removed from the y_i's by registration. On the other hand, MSE_{amp} is now a measure of pure amplitude variation to the extent that the registration has been successful. The decomposition does depend on the success of the registration step, however, since it is possible in principle for MSE_{phase} to be negative.

From this decomposition we can get a useful squared multiple correlation index of the proportion of the total variation due to phase:

$$R^2 = \frac{MSE_{phase}}{MSE_{total}}. \tag{8.6}$$

We applied used the decomposition to compare the unregistered acceleration curves with their landmark registered counterparts. Because the variation in the acceleration curves is far greater in the first few years than for the remaining years, which is the variation visible in Figure 8.1, we elected to use the decomposition over only the years from three to eighteen years. The function AmpPhaseDecomp returns a list with components MS.amp, MS.pha, RSQR and C. The commands

```
AmpPhasList = AmpPhaseDecomp(accffd, accregfdLM,
                             warpfd, c(3,18))
RSQR        = AmpPhasList$RSQR
```

after landmark registration of the growth acceleration curves yields the value $R^2 = 0.70$. That is, nearly 70% of the variation in acceleration over this period is due to phase.

On the other hand, if we use this decomposition to compare the landmark registered curves in Figure 8.1 with those for the continuously registered curves in Figure 8.4, we get the value −0.06. What does this mean? It means that "registered" is a rather fuzzy qualifier in the sense that we can define the registration process in different ways and get different answers. A careful comparison of the two figures might suggest that the landmark registration process has over–registered the pubertal growth spurt at the expense of earlier growth spurts visible in several of the curves. Or, alternatively, if our main concern is getting pubertal growth right, then the continuous registration process has deregistered the landmark–registered curves by about 6%.

8.6 Registering the Chinese Handwriting Data

The handwriting data discussed in Section 1.2.2 consisted of the writing of "statistics" in simplified Chinese 50 times. The average time of writing was six seconds, with the X-, Y- and Z-coordinates of the pen position being recorded 400 times per second. The handwriting involves 50 strokes, corresponding to about eight strokes per second, or 120 milliseconds per stroke. The processing of these data was done entirely in Matlab, and is too complex to describe in detail here.

The registration phase was carried out in two steps, as was the case for the growth data. In the first phase, three clear landmarks were visible in all curves in the vertical Z-coordinate corresponding to points where the pen was lifted from the paper. These were used in a preliminary landmark registration process for the Z-coordinate alone. The decomposition described above indicated that 66.6% of the variation in Z was due to phase. The warping functions were applied to the X- and Y-coordinates as well, and the decompositions indicated percentages of phase variation of 0% and 75%, respectively. This suggests that most of the phase variation in movement off the writing plane was associated with motion that was also vertical in the writing plane.

In a second registration phase, the scalar *tangential accelerations*,

$$TA_i(t) = \sqrt{D^2 X_i(t) + D^2 Y_i(t)},$$

of the tip of the pen along the writing path were registered using continuous registration. This corresponded to 48% of the variation in the landmark-registered tangential accelerations being due to phase. Figure 8.6 plots the tangential acceleration for all 50 replications before and after applying this two–stage registration procedure. After alignment, we see the remarkably small amount of amplitude variation in many of the acceleration peaks, and we also see how evenly spaced in time these peaks are. The pen hits acceleration of 30 meters/sec/sec, or three times the force of gravity. If sustained, this would launch a satellite into orbit in about seven minutes and put us in a plane's luggage rack if our seat belts were not fastened. It is also striking that near zero acceleration is found between these peaks.

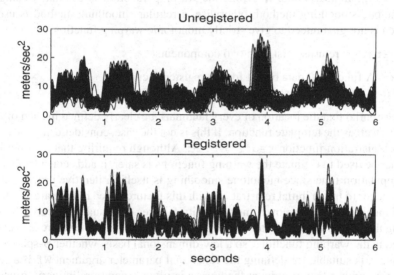

Fig. 8.6 The acceleration along the pen trajectory for all 50 replications of the script in Figure 1.9 before and after registration.

8.7 Details for Functions `landmarkreg` and `register.fd`

8.7.1 Function `landmarkreg`

The complete calling sequence for the R version is

```
landmarkreg(fdobj, ximarks, x0marks=xmeanmarks,
```

```
                           WfdPar, monwrd=FALSE)
```

The arguments are as follows:

fdobj   A functional data object containing the curves to be registered.

ximarks   A matrix containing the timings or argument values associated with the landmarks for the observations in fd to be registered. The number of rows N equals the number of observations and the number of columns NL equals the number of landmarks. These landmark times must be in the interior of the interval over which the functions are defined.

x0marks   A vector of times of landmarks for target curve. If not supplied, the mean of the landmark times in ximarks is used.

WfdPar   A functional parameter object defining the warping functions that transform time in order to register the curves.

monwrd   A logical value: if TRUE, the warping function is estimated using a monotone smoothing method; otherwise, a regular smoothing method is used, which is not guaranteed to give strictly monotonic warping functions.

Landmarkreg returns a list with two components:

fdreg   A functional data object for the registered curves.

warpfd   A functional data object for the warping functions.

It is essential that the location of every landmark be clearly defined in each of the curves as well as the template function. If this is not the case, consider using the continuous registration function register.fd. Although requiring that a monotone smoother be used to estimate the warping functions is safer, it adds considerably to the computation time since monotone smoothing is itself an iterative process. It is usually better to try an initial registration with this feature to see if there are any failures of monotonicity. Moreover, monotonicity failures can usually be cured by increasing the smoothing parameter defining WfdPar. Not much curvature is usually required in the warping functions, so a low-dimensional basis, whether B-splines or monomials, is suitable for defining the functional parameter argument WfdPar. A registration with a few prominent landmarks is often a good preliminary to using the more sophisticated but more lengthy process in register.fd.

8.7.2 Function register.fd

The complete calling sequence for the R version is

```
register.fd(y0fd=NULL, yfd=NULL,
            WfdParobj=c(Lfdobj=2, lambda=1),
            conv=1e-04, iterlim=20, dbglev=1,
            periodic=FALSE, crit=2)
```

y0fd   A functional data object defining the target for registration. If yfd is NULL and y0fd is a multivariate data object, then y0fd is assigned to yfd and y0fd is replaced by its mean. Alternatively, if `yfd` is a multivariate functional data object and `y0fd` is missing, `y0fd` is replaced by the mean of `y0fd`. Otherwise, `y0fd` must be a univariate functional data object taken as the target to which `yfd` is registered.

`yfd`   A multivariate functional data object defining the functions to be registered to target `y0fd`. If it is `NULL` and `y0fd` is a multivariate functional data object, `yfd` takes the value of `y0fd`.

`WfdParobj`   A functional parameter object for a single function. This is used as the initial value in the estimation of a function $W(t)$ that defines the warping function $h(t)$ that registers a particular curve. The object also contains information on a roughness penalty and smoothing parameter to control the roughness of $h(t)$. Alternatively, this can be a vector or a list with components named `Lfdobj` and `lambda`, which are passed as arguments to `fdPar` to create the functional parameter form of `WfdParobj` required by the rest of the `register.fd` algorithm. The default `Lfdobj` of 2 penalizes curvature, thereby preferring no warping of time, with `lambda` indicating the strength of that preference. A common alternative is `Lfdobj` = 3, penalizing the rate of change of curvature.

`conv`   A criterion for convergence of the iterations.

`iterlim`   A limit on the number of iterations.

`dbglev`   Either 0, 1, or 2. This controls the amount of information printed out on each iteration, with 0 implying no output, 1 intermediate output level, and 2 full output. (If this is run with output buffering, it may be necessary to turn off the output buffering to actually get the progress reports before the completion of computations.)

`periodic`   A logical variable: if `TRUE`, the functions are considered to be periodic, in which case a constant can be added to all argument values after they are warped.

`crit`   An integer that is either 1 or 2 that indicates the nature of the continuous registration criterion that is used. If 1, the criterion is least squares, and if 2, the criterion is the minimum eigenvalue of a cross-product matrix. In general, criterion 2 is to be preferred.

A named list of length 3 is returned containing the following components:

`regfd`   A functional data object containing the registered functions.

`Wfd`   A functional data object containing the functions $hW(t)$ that define the warping functions $h(t)$.

`shift`   If the functions are periodic, this is a vector of time shifts.

8.8 Some Things to Try

1. At the end of Chapter 7 we suggested a principal components analysis of the log of the first derivative of the growth curves. This was, of course, before registra-

tion. Now repeat this analysis for the registered growth curves, and compare the results. What about the impact of the pubertal growth spurt now?

2. Try applying continuous registration to the unregistered growth curves. You will see that a few curves are badly misaligned, indicating that there are limits to how well continuous registration works. What should we do with these misaligned curves? Could we try, for example, starting the continous registrations off with initial estimates of function Wfd set up from the landmark registered results?

3. Using only those girls whose curves are well registered by continuous registration, now use canonical correlation analysis to explore the covariation between the Wfd object returned by function register.fd and the Wfd object from the monotone smooth. Look for interesting ways in which the amplitude variation in growth is related to its the phase variation.

4. **Medfly Data**: In Section 7.7, we suggested applying principal components analysis to the medfly data. Here, we suggest you extend that analysis as follows:

   a. Perform a functional linear regression to predict the total lifespan of the fly from their egg laying. Choose a smoothing parameter by cross-validation, and plot the coefficient function along with confidence intervals.
   b. Conduct a permutation test for the significance of the regression. Calculate the R^2 for your regression.
   c. Compare the results of the functional linear regression with the linear regression on the principal component scores from your analysis in Section 7.7.

8.9 More to Read

The classic paper on the estimation of time warping functions is Sakoe and Chiba (1978), who used dynamic programming to estimate the warping function in a context where there was no need for the warping function to be smooth.

Landmark registration has been studied in depth by Kneip and Gasser (1992) and Gasser and Kneip (1995), who refer to a landmark as a *structural feature*, its location as a *structural point*, to the distribution of landmark locations along the t axis as *structural intensity*, and to the process of averaging a set of curves after registration as *structural averaging*. Their papers contain various technical details on the asymptotic behavior of landmark estimates and warping functions estimated from them. Another source of much information on the study of landmarks and their use in registration is Bookstein (1991).

The literature on continuous registration is evolving rapidly, but is still somewhat technical. Gervini and Gasser (2004) and Liu and Müller (2004) are recent papers that review the literature and discuss some theoretical issues.

Chapter 9
Functional Linear Models for Scalar Responses

This is the first of two chapters on the functional linear model. Here we have a dependent or response variable whose value is to be predicted or approximated on the basis of a set of independent or covariate variables, and at least one of these is functional in nature. The focus here is on linear models, or functional analogues of linear regression analysis. This chapter is confined to considering the prediction of a scalar response on the basis of one or more functional covariates, as well as possible scalar covariates.

Confidence intervals are developed for estimated regression functions in order to permit conclusions about where along the t axis a covariate plays a strong role in predicting a functional responses. The chapter also offers some permutation tests of hypotheses.

More broadly, we begin here the study of input/output systems. This and the next chapter lead in to Chapter 11, where the response is the derivative of the output from the system.

9.1 Functional Linear Regression with a Scalar Response

We have so far focused on representing a finite number of observations as a functional data object, which can theoretically be evaluated at an infinite number of points, and on graphically exploring the variation and covariation of populations of functions. However, we often need to model predictive relationships between different types of data, and here we expect that some of these data will be functional.

In classic linear regression, predictive models are often of the form

$$y_i = \sum_{j=0}^{p} x_{ij}\beta_j + \varepsilon_i, \quad i = 1,\dots,N \tag{9.1}$$

that model the relationship between a *response* y_i and *covariates* x_{ij} as a *linear* structure. The dummy covariate with $j = 0$, which has the value one for all i, is

J.O. Ramsay et al., *Functional Data Analysis with R and MATLAB*, Use R, DOI: 10.1007/978-0-387-98185-7_9, © Springer Science + Business Media, LLC 2009

usually included because origin of the response variable and/or one or more of the independent variables can be arbitrary, and β_0 codes the constant needed to allow for this. It is often called the *intercept* term.

The term ε_i allows for sources of variation considered extraneous, such as measurement error, unimportant additional causal factors, sources of nonlinearity and so forth, all swept under the statistical rug called *error*. The ε_i are assumed to add to the prediction of the response, and are usually considered to be independently and identically distributed.

In this chapter we replace at least one of the p observed scalar covariate variables on the right side of the classic equation by a functional covariate. To simplify the exposition, though, we will describe a model consisting of a single functional independent variable plus an intercept term.

9.2 A Scalar Response Model for Log Annual Precipitation

Our test-bed problem in this section is to predict the logarithm of annual precipitation for 35 Canadian weather stations from their temperature profiles. The response in this case is, in terms of the fda package in R,

```
annualprec = log10(apply(daily$precav,2,sum))
```

We want to use as the predictor variable the complete temperature profile as well as a constant intercept. These two covariates can be stored in a list of length 2, or in Matlab as a cell array. Here we set up a functional data object for the 35 temperature profiles, called tempfd. To keep things simple and the computation rapid, we will use 65 basis functions without a roughness penalty. This number of basis functions has been found to be adequate for most purposes, and can, for example, capture the ripples observed in early spring in many weather stations.

```
tempbasis =create.fourier.basis(c(0,365),65)
tempSmooth=smooth.basis(day.5,daily$tempav,tempbasis)
tempfd    =tempSmooth$fd
```

9.3 Setting Up the Functional Linear Model

But what can we do when the vector of covariate observations $\mathbf{x}_i = (x_{i1}, \ldots, x_{ip})$ in (9.1) is replaced by a function $x_i(t)$? A first idea might be to discretize each of the N functional covariates $x_i(t)$ by choosing a set of times $t_1, \ldots, t_q$ and consider fitting the model

$$y_i = \alpha_0 + \sum_{j=1}^{q} x_i(t_j)\beta_j + \varepsilon_i.$$

But which times t_j are important, given that we must have $q < N$?

If we choose a finer and finer mesh of times, the summation approaches an integral equation:

$$y_i = \alpha_0 + \int x_i(t)\beta(t)\mathrm{d}t + \varepsilon_i. \tag{9.2}$$

We now have a finite number N of observations with which to determine the infinite-dimensional $\beta(t)$. This is an impossible problem: it is almost always possible to find a $\beta(t)$ so that (9.2) is satisfied with $\varepsilon_i = 0$. More importantly, there are always an infinite number of possible regression coefficient functions $\beta(t)$ that will produce exactly the same predictions $\hat{y}_i$. Even if we expand each functional covariate in terms of a limited number of basis functions, it is entirely possible that the total number of basis functions will exceed or at least approach N.

9.4 Three Estimates of the Regression Coefficient Predicting Annual Precipitation

Three strategies have been developed to deal with this underdetermination issue. The first two redefine the problem using a basis coefficient expansion of β:

$$\beta(t) = \sum_k^K c_k \phi_k(t) = \mathbf{c}'\boldsymbol{\phi}(t). \tag{9.3}$$

The third replaces the potentially high-dimensional covariate functions by a low-dimensional approximation using principal components analysis. The first two approaches will be illustrated using function fRegress. Function fRegress in R and Matlab requires at least three arguments:

yfdPar This object contains the response variable. It can be a functional parameter object, a functional data object, or simply a vector of N scalar responses. In this chapter we restrict ourselves to the scalar response situation.

xfdlist This object contains all the functional and scalar covariate functions used to predict the response using the linear model. Each covariate is an element or component in a list object in R or an entry in a cell array in Matlab.

betalist This is a list object in R or a cell array in Matlab of the same length as the second argument; it specifies the functional regression coefficient objects. Because it is possible that any or all of them can be subject to a roughness penalty, they are all in principle functional parameter objects, although fRegress will by default convert both functional data objects and basis objects into functional parameter objects.

Here we store the two functional data covariates required for predicting log annual precipitation in a list of length two, which we here call templist, to be used for the argument xfdlist.

```
templist       = vector("list",2)
templist[[1]]  = rep(1,35)
templist[[2]]  = tempfd
```

Notice that the intercept term is here set up as a constant function with 35 replications.

9.4.1 Low-Dimensional Regression Coefficient Function β

The simplest strategy for estimating β is just to keep the dimensionality K of β in (9.3) small relative to N. In our test-bed expansion, we will work with five Fourier basis functions for the regression coefficient β multiplying the temperature profiles; we will also use a constant function for α, the multiplier of the constant intercept covariate set up above.

```
conbasis       = create.constant.basis(c(0,365))
betabasis      = create.fourier.basis(c(0,365),5)
betalist       = vector("list",2)
betalist[[1]]  = conbasis
betalist[[2]]  = betabasis
```

Now we can call function fRegress, which returns various results in a list object that we call fRegressList.

```
fRegressList = fRegress(annualprec,templist,betalist)
```

The command names(fRegressList) reveals a component betaestlist containing the estimated regression coefficient functions. Each of these is a functional parameter object. We can plot the estimate of the regression function for the temperature profiles with the commands

```
betaestlist = fRegressList$betaestlist
tempbetafd  = betaestlist[[2]]$fd
plot(tempbetafd, xlab="Day",
     ylab="Beta for temperature")
```

Figure 9.1 shows the result. The intercept term can be obtained from coef(betaestlist[[1]]); its value in this case is 0.0095. We will defer commenting on these estimates until we consider the next more sophisticated strategy.

We need to assess the quality of this fit. First, let us extract the fitted values defined by this model and compute the residuals. We will also compute error sums of squares associated with the fit as well as for the fit using only a constant or intercept.

```
annualprechat1 = fRegressList$yhatfdobj
annualprecres1 = annualprec - annualprechat1
SSE1.1 = sum(annualprecres1^2)
SSE0 = sum((annualprec - mean(annualprec))^2)
```

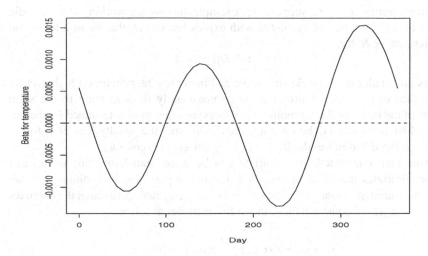

Fig. 9.1 Estimated $\beta(t)$ for predicting log annual precipitation from average daily temperature using five Fourier basis functions.

We can now compute the squared multiple correlation and the usual F-ratio for comparing these two fits.

```
RSQ1    = (SSE0-SSE1.1)/SSE0
Fratio1 = ((SSE0-SSE1)/5)/(SSE1/29)
```

The squared multiple correlation is 0.80, and the corresponding F-ratio with 5 and 29 degrees of freedom is 22.6, suggesting a fit to the data that is far better than we would expect by chance.

9.4.2 Coefficient β Estimate Using a Roughness Penalty

There are two ways to obtain a smooth fit. The simplest is to use a low-dimensional basis for $\beta(t)$. However, we can get more direct control over what we mean by "smooth" by using a roughness penalty. The combination of a high-dimensional basis with a roughness penalty reduces the possibilities that either (*a*) important features are missed or (*b*) extraneous features are forced into the image by using a basis set that is too small for the application.

Suppose, for example, that we fit (9.2) by minimizing the penalized sum of squares

$$\text{PENSSE}_\lambda(\alpha_0, \beta) = \sum [y_i - \alpha_0 - \int x_i(t)\beta(t)\mathrm{d}t]^2 + \lambda \int [L\beta(t)]^2 \mathrm{d}t. \qquad (9.4)$$

This allows us to shrink variation in β as close as we wish to the solution of the *differential equation* $L\beta = 0$. Suppose, for example, that we are working with periodic data with a known period. As noted with expression (5.11), the use of a harmonic acceleration operator,

$$L\beta = (\omega^2)D\beta + D^3\beta,$$

places no penalty on a simple sine wave and increases the penalty on higher-order harmonics in a Fourier approximation approximately in proportion to the sixth power of the order of the harmonic. (In this expression, ω is determined by the period, which is assumed to be known.) Thus, increasing the penalty λ in (9.4) forces β to look more and more like $\beta(t) = c_1 + c_2 \sin(\omega t) + c_3 \cos(\omega t)$.

More than one functional covariate can be incorporated into this model and scalar covariates may also be included. Let us suppose that in addition to y_i we have measured p scalar covariates $\mathbf{z}_i = (z_{i1},\ldots,z_{ip})$ and q functional covariates $x_{i1}(t),\ldots,x_{iq}(t)$. We can put these into a linear model as follows

$$y_i = \alpha_0 + \mathbf{z}_i'\alpha + \sum_{j=1}^{q} \int x_{ij}(t)\beta_j(t)\mathrm{d}t + \varepsilon_i, \qquad (9.5)$$

where $\mathbf{z}_i$ is the $p - vector$ of scalar covariates. A separate smoothing penalty may be employed for each of the $\beta_j(t), j = 1,\ldots,q$.

Using (9.5), we can define a least-squares estimate as follows. We define $\mathbf{Z}$ by:

$$\mathbf{Z} = \begin{bmatrix} \mathbf{z}_1' & \int x_{11}(t)\Phi_1(t)dt & \cdots & \int x_{1q}(t)\Phi_q(t) \\ \vdots & & & \vdots \\ \mathbf{z}_n' & \int x_{n1}(t)\Phi_1(t)dt & \cdots & \int x_{nq}(t)\Phi_q(t) \end{bmatrix}.$$

Here Φ_k is the basis expansion used to represent $\beta_k(t)$. We also define a penalty matrix

$$\mathbf{R}(\lambda) = \begin{bmatrix} 0 & \cdots & \cdots & \cdots \\ 0 & \lambda_1 R_1 & \cdots & 0 \\ \vdots & \vdots & \ddots & \vdots \\ 0 & \cdots & \lambda_q R_q \end{bmatrix} \qquad (9.6)$$

where R_k is the penalty matrix associated with the smoothing penalty for β_k and λ_k is the corresponding smoothing parameter. With these objects, we can define

$$\hat{\mathbf{b}} = \left(\mathbf{Z}'\mathbf{Z} + \mathbf{R}(\lambda)\right)^{-1}\mathbf{Z}'\mathbf{y}$$

to hold the vector of estimated coefficients $\hat{\alpha}$ along with the coefficients defining each estimated coefficient function $\hat{\beta}_k(t)$ estimated by penalized least squares. These are then extracted to form the appropriate functional data objects.

Now let us apply this approach to predicting the log annual precipitations. First, we set up a harmonic acceleration operator, as we did already in Chapter 5.

```
Lcoef = c(0,(2*pi/365)^2,0)
```

```
harmaccelLfd = vec2Lfd(Lcoef, c(0,365))
```

Now we replace our previous choice of basis for defining the β estimate by a functional parameter object that incorporates both this roughness penalty and a level of smoothing:

```
betabasis = create.fourier.basis(c(0, 365), 35)
lambda    = 10^12.5
betafdPar = fdPar(betabasis, harmaccelLfd, lambda)
betalist[[2]] = betafdPar
```

These now allow us to invoke fRegress to return the estimated functional coefficients and predicted values:

```
annPrecTemp    = fRegress(annualprec, templist,
                          betalist)
betaestlist2   = annPrecTemp$betaestlist
annualprechat2 = annPrecTemp$yhatfdobj
```

The command print(fRegressList$df) indicates the degrees of freedom for this model, including the intercept, is 4.7, somewhat below the value of 6 that we used for the simple model above.

Now we compute the usual R^2 and F-ratio statistics to assess the improvement in fit achieved by including temperature as a covariate.

```
SSE1.2   = sum((annualprec-annualprechat2)^2)
RSQ2     = (SSE0 - SSE1.2)/SSE0
Fratio2 = ((SSE0-SSE1.2)/3.7)/(SSE1/30.3)
```

The squared multiple correlation is now 0.75, a small drop from the value for the simple model, due partly to using fewer degrees of freedom. The F-ratio is 25.1 with 3.7 and 30.3 degrees of freedom, and is even more significant than for the simple model. The reader should note that because a smoothing penalty has been used, the F-distribution only represents an approximation to the null distribution for this model. Figure 9.2 compares predicted and observed values of log annual precipitation. Figure 9.3 plots the coefficient $\beta(t)$ along with the confidence intervals derived below. Comparing this version with that in Figure 9.1 shows why the roughness penalty approach is to be preferred over the fixed low-dimension strategy; now we see that only the autumn months really matter in defining the relationship, and that the substantial oscillations over other parts of the year in Figure 9.1 are actually extraneous.

To complete the picture, we should ask whether we could do just as well with a constant value for β. Here we use the constant basis, run fRegress, and redo the comparison using this fit as a benchmark. The degrees of freedom for this model is now 2.

```
betalist[[2]] = fdPar(conbasis)
fRegressList   = fRegress(annualprec, templist,
                          betalist)
betaestlist    = fRegressList$betaestlist
```

Now we compute the test statics for comparing these models.

```
annualprechat = fRegressList$yhatfdobj
SSE1          = sum((annualprec-annualprechat)^2)
RSQ           = (SSE0 - SSE1)/SSE0
Fratio        = ((SSE0-SSE1)/1)/(SSE1/33)
```

We find that $R^2 = 0.49$ and $F = 31.3$ for 1 and 33 degrees of freedom, so that the contribution of our model is also important relative to this benchmark. That is, functional linear regression is the right choice here.

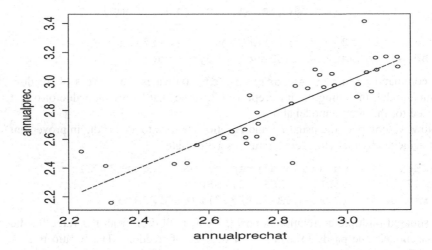

Fig. 9.2 Observed log annual precipitation values plotted against values predicted by functional linear regression on temperature curves using a roughness penalty.

9.4.3 Choosing Smoothing Parameters

How did we come up with $\lambda = 10^{12.5}$ for the smoothing parameter in this analysis? Although smoothing parameters λ_j can certainly be chosen subjectively, we can also consider *cross-validation* as a way of using the data to define smoothing level. To define a cross-validation score, we let $\alpha_\lambda^{(-i)}$ and $\beta_\lambda^{(-i)}$ be the estimated regression parameters estimated without the ith observation. The cross-validation score is then

$$\mathrm{CV}(\lambda) = \sum_{i=1}^{N} \left[y_i - \alpha_\lambda^{(-i)} - \int x_i(t)\beta_\lambda^{(-i)} \mathrm{d}t \right]^2. \tag{9.7}$$

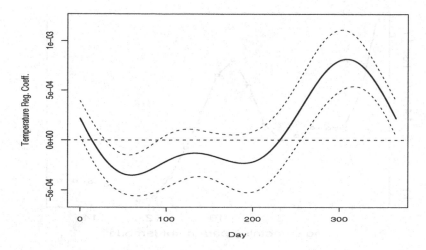

Fig. 9.3 Estimate $\beta(t)$ for predicting log annual precipitation from average daily temperature with a harmonic acceleration penalty and smoothing parameter set to $10^{12.5}$. The dashed lines indicate pointwise 95% confidence limits for values of $\beta(t)$.

Observing that

$$\hat{\mathbf{y}} = \mathbf{Z}\left(\mathbf{Z}'\mathbf{Z} + R(\lambda)\right)^{-1}\mathbf{Z}'\mathbf{y} = \mathbf{Hy}$$

standard calculations give us that

$$\mathrm{CV}(\lambda) = \sum_{i=1}^{N}\left(\frac{y_i - \hat{y}_i}{1 - H_{ii}}\right)^2. \qquad (9.8)$$

We can similarly define a generalized cross-validation score:

$$\mathrm{GCV}(\lambda) = \frac{\sum_{i=1}^{n}(y_i - \hat{y}_i)^2}{(n - \mathrm{Tr}(H))^2}. \qquad (9.9)$$

These quantities are returned by `fRegress` for scalar responses only. This $\mathrm{GCV}(\lambda)$ was discussed (in different notation) in Section 5.2.5. For a comparison of CV and GCV including reference to more literature, see Section 21.3.4, p. 368, in Ramsay and Silverman (2005).

The following code generates the data plotted in Figure 9.4.

```
loglam = seq(5,15,0.5)
nlam   = length(loglam)
SSE.CV = matrix(0,nlam,1)
for (ilam in 1:nlam) {
  lambda              = 10^loglam[ilam]
```

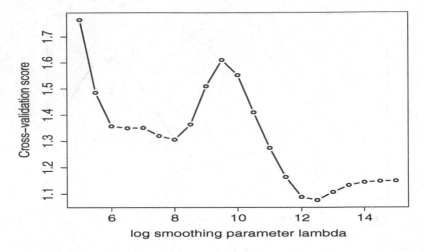

Fig. 9.4 Cross-validation scores CV(λ) for fitting log annual precipitation by daily temperature profile, with a penalty on the harmonic acceleration of $\beta(t)$.

```
      betalisti           = betalist
      betafdPar2          = betalisti[[2]]
      betafdPar2$lambda   = lambda
      betalisti[[2]]      = betafdPar2
      fRegi = fRegress.CV(annualprec, templist,
                                    betalisti)
      SSE.CV[ilam] = fRegi$SSE.CV
   }
```

9.4.4 Confidence Intervals

Once we have conducted a functional linear regression, we want to measure the precision to which we have estimated each of the $\hat{\beta}_j(t)$. This can be done in the same manner as confidence intervals for probes in smoothing. Under the usual independence assumption, the ε_i are independently normally distributed around zero with variance σ_e^2. The covariance of ε is then

$$\Sigma = \sigma_e^2 I.$$

Following a δ-method calculation, the sampling variance of the estimated parameter vector $\hat{\mathbf{b}}$ is

$$\text{Var}\left[\hat{\mathbf{b}}\right] = \left(\mathbf{Z}'\mathbf{Z} + \mathbf{R}(\lambda)\right)^{-1}\mathbf{Z}'\Sigma\mathbf{Z}\left(\mathbf{Z}'\mathbf{Z} + R\right)^{-1}.$$

Naturally, any more general estimate of Σ, allowing correlation between the errors, can be used here.

We can now obtain confidence intervals for each of the $\beta_j(t)$. To do this, we need an estimate of σ_e^2. This can be obtained from the residuals. The following code does the trick in R:

```
resid    = annualprec - annualprechat
SigmaE.= sum(resid^2)/(35-fRegressList$df)
SigmaE = SigmaE.*diag(rep(1,35))
y2cMap = tempSmooth$y2cMap
stderrList = fRegress.stderr(fRegressList, y2cMap,
                             SigmaE)
```

Here the second argument to `fRegress.stderr` is a place-holder for a projection matrix that will be used for functional responses (see Chapter 10). We can then plot the coefficient function $\beta(t)$ along with plus and minus two times its standard error to obtain the approximate confidence bounds in Figure 9.3:

```
betafdPar        = betaestlist[[2]]
betafd           = betafdPar$fd
betastderrList   = stderrList$betastderrlist
betastderrfd     = betastderrList[[2]]
plot(betafd, xlab="Day",
             ylab="Temperature Reg. Coeff.",
             ylim=c(-6e-4,1.2e-03), lwd=2)
lines(betafd+2*betastderrfd, lty=2, lwd=1)
lines(betafd-2*betastderrfd, lty=2, lwd=1)
```

We note that, like the confidence intervals that we derived for probes, these intervals are given pointwise and do not take account of bias or of the choice of smoothing parameters. In order to provide tests for the overall effectiveness of the regression we resort to permutation tests described in Section 9.5 below.

9.4.5 Scalar Response Models by Functional Principal Components

A third alternative for functional linear regression with a scalar response is to regress $\mathbf{y}$ on the principal component scores for functional covariate. The use of principal components analysis in multiple linear regression is a standard technique:

1. Perform a principal components analysis on the covariate matrix X and derive the principal components scores f_{ij} for each observation i on each principal component j.
2. Regress the response y_i on the principal component scores c_{ij}.

We often observe that we need only the first few principal component scores, thereby considerably improving the stability of the estimate by increasing the degrees of freedom for error.

In functional linear regression, we consider the scores resulting from a functional principal components analysis of the temperature curves conducted in Chapter 7. We can write

$$x_i(t) = \bar{x}(t) + \sum_{j>=0} c_{ij}\xi_j(t).$$

Regressing on the principal component scores gives us the following model:

$$y_i = \beta_0 + \sum c_{ij}\beta_j + \varepsilon_i. \tag{9.10}$$

Now we recall that $c_{ij} = \int \xi_j(t)(x_i(t) - \bar{x}(t))\mathrm{d}t$. If we substitute this in (9.10), we can see that

$$y_i = \beta_0 + \int \sum \beta_j\xi_j(t)(x_{ij}(t) - \bar{x}(t))\mathrm{d}t + \varepsilon_i.$$

This gives us

$$\beta(t) = \sum \beta_j\xi_j(t).$$

Thus, (9.10) expresses exactly the same relationship as (9.2) when we absorb the mean function into the intercept:

$$\tilde{\beta}_0 = \beta_0 - \int \beta(t)\bar{x}(t)dt.$$

The following code carries out this idea for the annual cycles in daily temperatures at 35 Canadian weather stations. First we resmooth the data using a saturated basis with a roughness penalty. This represents rather more smoothing than in the earlier version of tempfd that did not use a roughness penalty.

```
daybasis365=create.fourier.basis(c(0, 365), 365)
lambda     =1e6
tempfdPar  =fdPar(daybasis365, harmaccelLfd, lambda)
tempfd     =smooth.basis(day.5, daily$tempav,
                         tempfdPar)$fd
```

Next we perform the principal components analysis, again using a roughness penalty.

```
lambda    = 1e0
tempfdPar = fdPar(daybasis365, harmaccelLfd, lambda)
temppca   = pca.fd(tempfd, 4, tempfdPar)
harmonics = temppca$harmonics
```

Approximate pointwise standard errors can now be constructed out of the covariance matrix of the β_j:

$$\text{var}[\hat{\beta}(t)] = [\xi_1(t) \; \dots \; \xi_k(t)]\text{Var}[\beta] \begin{bmatrix} \xi_1(t) \\ \vdots \\ \xi_k(t) \end{bmatrix}.$$

Since the coefficients are orthogonal, the covariance of the β_j is diagonal and can be extracted from the standard errors reported by `lm`. When smoothed principal components are used, however, this orthogonality no longer holds and the full covariance must be used.

The final step is to do the linear model using principal component scores and to construct the corresponding functional data objects for the regression functions.

```
pcamodel = lm(annualprec~temppca$scores)
pcacoefs = summary(pcamodel)$coef
betafd   = pcacoefs[2,1]*harmonics[1] +
           pcacoefs[3,1]*harmonics[2] +
           pcacoefs[4,1]*harmonics[3]
coefvar  = pcacoefs[,2]^2
betavar  = coefvar[2]*harmonics[1]^2 +
           coefvar[3]*harmonics[2]^2 +
           coefvar[4]*harmonics[3]^2
```

The quantities resulting from the code below are plotted in Figure 9.5. In this case the R-squared statistic is similar to the previous analysis at 0.72.

```
plot(betafd, xlab="Day", ylab="Regression Coef.",
        ylim=c(-6e-4,1.2e-03), lwd=2)
lines(betafd+2*sqrt(betavar), lty=2, lwd=1)
lines(betafd-2*sqrt(betavar), lty=2, lwd=1)
```

Functional linear regression by functional principal components has been studied extensively. Yao et al. (2005) observes that instead of presmoothing the data, we can estimate the covariance surface directly by a two-dimensional smooth and use this to derive the fPCA. From here the principal component scores can be calculated by fitting the principal component functions to the data by least squares. This can be advantageous when some curves are sparsely observed.

9.5 Statistical Tests

So far, our tools have concentrated on *exploratory* analysis. We have developed approximate pointwise confidence intervals for functional coefficients. However, we have not attempted to formalize these into test statistics. Hypothesis tests provide a formal criterion for judging whether a scientific hypothesis is valid. They also perform the useful function of allowing us to assess "What would the results look like if there really were no relationship in the data?"

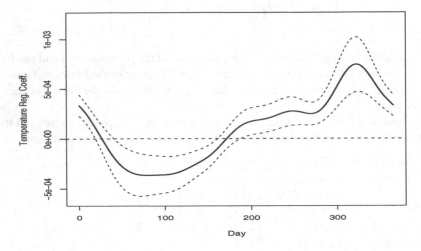

Fig. 9.5 Estimate $\beta(t)$ for predicting log annual precipitation from average daily temperature using scores from the first three functional principal components of temperature. The dashed lines indicate pointwise 95% confidence limits for values of $\beta(t)$.

Because of the nature of functional statistics, it is difficult to attempt to derive a theoretical null distribution for any given test statistic since we would need to account for selecting a smoothing parameter as well as the smoothing itself. Instead, the package employs a permutation test methodology. This involves constructing a null distribution from the observed data directly. If there is no relationship between the response and the covariates, it should make no difference if we randomly rearrange the way they are paired. To see what a result might look in this case, we can simply perform the experiment of rearranging the vector of responses while keeping the covariates in the same order and trying to fit the model again. The advantage of this is that we no longer need to rely on distributional assumptions. The disadvantage is that we cannot test for the significance of an individual covariate among many.

In order to turn this idea into a formalized statistical procedure, we need a way to determine whether the result we get from the observed data is different from what is obtained by rearranging the response vector. We do this in the classic manner, by deciding on a test statistic that measures the strength of the predictive relationship in our model. We now have a single number which we can compare to the distribution that is obtained when we calculate the same statistic with a randomly permuted response. If the observed test statistic is in the tail of this distribution, we conclude that there is a relationship between the response and covariates.

In our case, we compute an F statistic for the regression

$$F = \frac{\text{Var}[\hat{y}]}{\frac{1}{n}\sum(y_i - \hat{y}_i)^2}$$

where $\hat{y}$ is the vector of predicted responses. This statistic varies from the classic F statistic in the manner in which it normalizes the numerator and denominator sums of squares. The statistic is calculated several hundred times using a different random permutation each time. The p value for the test can then be calculated by counting the proportion of permutation F values that are larger than the F statistic for the observed pairing.

The following code implements this procedure for the Canadian weather example:

```
F.res = Fperm.fd(annualprec, templist, betalist)
```

Here the observed F statistic (stored in F.res$Fobs) is 3.03, while the 95th quartile of the permutation distribution (F.res$qval) is 0.27, giving strong evidence for the effect.

9.6 Some Things to Try

1. **Medfly Data**: We reconsider the medfly data described in Section 7.7.

   a. Perform a functional linear regression to predict the total lifespan of the fly from their egg laying. Choose a smoothing parameter by cross-validation and plot the coefficient function with confidence intervals.
   b. What is the R^2 of your fit to these data? How does this compare with that for the principal components regression you tried earlier?
   c. Construct confidence intervals for the coefficient function obtained using principal components regression. How do these compare to that for your estimates found using fRegress? Experiment with increasing the smoothing parameter in fRegress and the number of components for principal components regression.
   d. Conduct a permutation test for the significance of the regression. Calculate the R^2 for your regression.

2. **Tecator Data**: The Tecator data available from

   http://lib.stat.cmu.edu/datasets/tecator

   provides an example of functional data in which the domain of the function is not time. Instead, we observe the spectra of meat samples from which we would like to determine a number of chemical components. In particular, the moisture content of the meat is of interest.

   a. Represent the spectra using a reasonable basis and smoothing penalty.

   b. Experiment with functional linear regression using these spectra as covariates for the regression model. Plot the coefficient function along with confidence intervals. What is the R^2 for your regression?

   c. Try using the derivative of the spectrum to predict moisture content. What happens if you use both the derivative and the original spectrum?

3. **Diagnostics**: Residual diagnostics for functional linear regression is a largely unexplored area. Here are some suggestions for checking regression assumptions for one of the models suggested above.

   a. Residual by predicted plots can still be constructed, as can QQ plots for residuals. Do these tell you anything about the data?

   b. In linear regression, we look for curvature in a model by plotting residuals against covariates. In functional linear regression, we would need to plot residuals against the predictor value at each time t. Experiment with doing this at a fine grid of t. Alternatively, you can plot these as lines in three-dimensional space using the `lattice` or `rgl` package.

   c. Do any points concern you as exhibiting undue influence? Consider removing them and measure the effect on your model. One way to get an idea of influence is the integrated squared difference in the $\beta_i(t)$ coefficients. You can calculate this using the `inprod` function.

9.7 More to Read

Functional linear regression for scalar responses has a large associated literature. Models based on functional principal components analysis are found in Cardot et al. (1999), Cardot et al. (2003a) and Yao et al. (2005). Tests for no effect are developed in Cardot et al. (2004), Cardot et al. (2003b) and Delsol et al. (2008). More recent work by James et al. (2009) has focused on using absolute value penalties to insist that $\beta(t)$ be zero or exactly linear over large regions.

   Escabias et al. (2004) and James (2002) look at the larger problem of how to adapt the generalized linear model to the presence of a functional predictor variable. Müller and Stadtmüller (2005) also investigate what they call the *generalized functional linear model*. James and Hastie (2001) consider linear discriminant analysis where at least one of the independent variables used for prediction is a function and where the curves are irregularly sampled.

Chapter 10
Linear Models for Functional Responses

In this second chapter on the functional linear model, the dependent or response variable is functional. We first consider a situation in which all of the independent variables are scalar and in particular look at two functional analyses of variance.

When one or more of the independent variables is also function, we have two possible classes of linear models. The simpler case is called `concurrent`, where the value of the response variable $y(t)$ is predicted solely by the values of one or more functional covariates at the same time t. The more general case where functional variables contribute to the prediction for all possible time values s is briefly reviewed.

10.1 Functional Responses and an Analysis of Variance Model

While we often find functional covariates associated with scalar responses, there are also cases where the interest lies in the prediction of a functional response. We begin this chapter with two examples of *functional analysis of variance* (fANOVA), where variation in a functional response is decomposed into functional effects through the use of a scalar design matrix $\mathbf{Z}$. That is, in both of these examples, the covariates are all scalar.

10.1.1 Climate Region Effects on Temperature

In the Canadian weather data, for example, we can divide the weather stations into four distinct groups: Atlantic, Pacific, Prairie and Arctic. It may be interesting to know the effect of geographic location on the shape of the temperature curves. That is, we have a model of the form

J.O. Ramsay et al., *Functional Data Analysis with R and MATLAB*, Use R,
DOI: 10.1007/978-0-387-98185-7_10,
© Springer Science + Business Media, LLC 2009

$$y_i(t) = \beta_0(t) + \sum_{j=1}^{4} x_{ij}\beta_j(t) + \varepsilon_i(t) \qquad (10.1)$$

where $y_i(t)$ is a *functional response*. In this case, the values of x_{ij} are either 0 or 1. If the 35 by 5 matrix $\mathbf{Z}$ contains these values, then the first column has all entries equal to 1, which codes the contribution of the Canadian mean temperature; the remaining four columns contain 1 if that weather station is in the corresponding climate zone and 0 otherwise. In order to identify the specific effects of the four climate zones, we have to add the constraint

$$\sum_{j=1}^{4} \beta_j(t) = 0 \text{ for all } t. \qquad (10.2)$$

There are a number of methods of imposing this constraint. In this example we will do this by adding the above equation as an additional 36th "observation" for which $y_{36}(t) = 0$.

We first create a list containing five indicator variables for the intercept term and each of the regions. In this setup, the intercept term is effectively the Canadian mean temperature curve, and each of the remaining regression coefficients is the perturbation of the Canadian mean required to fit a region's mean temperature. These indicator variables are stored in the List object regionList.

```
regions       = unique(CanadianWeather$region)
p             = length(regions) + 1
regionList    = vector("list", p)
regionList[[1]] = c(rep(1,35),0)
for (j in 2:p) {
  xj = CanadianWeather$region == regions[j-1]
  regionList[[j]] = c(xj,1)
}
```

The next step is to augment the temperature functional data object by a 36th observation that takes only zero values as required by (10.2).

```
coef      = tempfd$coef
coef36    = cbind(coef,matrix(0,65,1))
temp36fd  = fd(coef36,tempbasis,tempfd$fdnames)
```

We now create functional parameter objects for each of the coefficient functions, using 11 Fourier basis functions for each.

```
betabasis = create.fourier.basis(c(0, 365), 11)
betafdPar = fdPar(betabasis)
betaList  = vector("list",p)
for (j in 1:p) betaList[[j]] = betafdPar
```

Now call fRegress, extract the coefficients and plot them, along with the predicted curves for the regions.

```
fRegressList = fRegress(temp36fd, regionList,
                        betaList)
betaestList   = fRegressList$betaestlist
regionFit     = fRegressList$yhatfd
regions       = c("Canada", regions)
par(mfrow=c(2,3),cex=1)
for (j in 1:p) plot(betaestList[[j]]$fd, lwd=2,
        xlab="Day (July 1 to June 30)",
        ylab="", main=regions[j])
plot(regionFit, lwd=2, col=1, lty=1,
        xlab="Day", ylab="",
        main="Prediction")
```

The five regression coefficients are shown in Figure 10.1; the final panel shows the predicted mean temperature curves for each of the four regions.

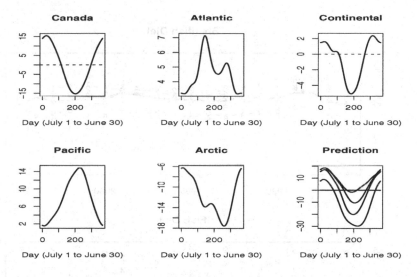

Fig. 10.1 The regression coefficients estimated for predicting temperature from climate region. The first panel is the intercept coefficient, corresponding to the Canadian mean temperature. The last panel contains the predicted mean temperatures for the four regions.

10.1.2 Trends in Seabird Populations on Kodiak Island

Winter abundance of 13 seabird species has been monitored annually in a number of bays on Kodiak Island, Alaska, since 1979 (Zwiefelhofer et al., 2008). Since 1986

these marine surveys have been conducted by the Kodiak National Wildlife Refuge
using a standard protocol revisiting a fixed set of transects in each bay. The bird
counts analyzed here are from two bays, Uyak and Uganik, both influenced by the
waters of the Shelikof Strait. We focus on results from 1991 to 2005, less 1998 when
the boat was in dry dock. We want to investigate potential differences in time trends
of bird species that primarily eat fish compared to those that primarily eat shellfish
and mollusks.

Figure 10.2 shows the base 10 logarithms of counts of the 13 species averaged
over transects and sites, and separated according to diet. It is obvious that there
are substantial differences in abundances over species that are consistent across the
years of observation, and there is more variation in abundance among the fish-eating
birds. The two mean functions suggest a slight tendency for the fish-eating birds to
be increasing in abundance relative to what we see for shellfish and mollusk eaters,
although this may be due to the sharply increasing trend for one fish-eating species.
We will use fRegress to see how substantial this difference is.

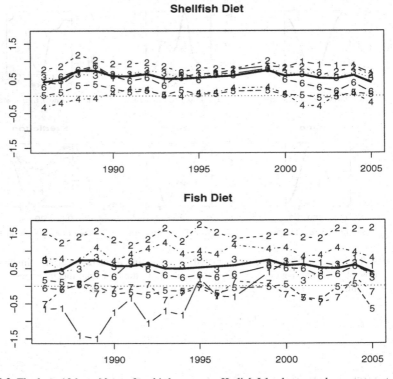

Fig. 10.2 The base 10 logarithms of seabird counts on Kodiak Island averaged over transects and
sites. The top panel shows counts for birds who eat shellfish and mollusks, and the bottom shows
counts for fish-eating birds. In each panel, the mean count taken across birds is shown as a heavy
solid line.

We elected to fit the count data exactly using a polygonal basis, since we were less interested in estimating smooth trends for each species than we were in estimating the functional diet effect. By interpolating the data in this way, we were sure to retain all the information in the original data. The following two commands set up the polygonal basis and fit the data in the 19 by 26 matrix `logCounts2`, and `yearCode = c(1:12, 14:20)` (because no data were collected for 1986, year 13). In this matrix, the first 13 columns are for the Uganik site and the remaining 13 for the Uyak site, and each row corresponds to a year of observation.

```
birdbasis  =  create.polygonal.basis(yearCode)
birdList   =  smooth.basis(yearCode,logCounts2,birdbasis)
birdfd     =  birdList$fd
```

We analyze the time trend in the log mean counts as affected by the type of diet, with bird species nested within the diet factor. The model that we consider is

$$y_{ijk}(t) = \mu(t) + (-1)^i \alpha(t) + \beta_{ij}(t) + \varepsilon_{ijk}(t) \qquad (10.3)$$

where $i = 1, 2$ indexes food groups, $j = 1, \ldots, n_i$ indexes birds within a food group, and $k = 1, 2$ indexes sites. The functional parameter $\mu(t)$ is the intercept or grand mean indicating the average trend for all birds. Parameter $\alpha(t)$ will indicate the time trend of the mean *difference* between the shellfish/mollusk-eating birds and the fish-eating birds, and multiplies a scalar covariate taking two values: 1 if an observation is for a shellfish/mollusk eater, and -1 otherwise. The 13 parameters $\beta_{ij}(t)$ are time trends for each bird, but represent deviations from $\mu(t)$ that are specific to a food group. This is achieved by imposing two constraint equations: $\sum_j \beta_{1j}(t) = 0$ and $\sum_j \beta_{2j}(t) = 0$ for all values of t. In each of these summations, index j takes values only within the corresponding food group. Finally, $\varepsilon_{ijk}(t)$ is the inevitable residual function required to balance each equation. The total number of equations is 28, being two blocks of 13 species plus one constraint for each food groups. There are two blocks, one for each bay or observation site.

To set up this model, we first define a variable for the diet effect, containing values 1 for a shellfish eater and -1 for a fish eater. This covariate effectively indicates the difference between being a shellfish eater and a fish eater.

```
foodindex = c(1,2,5,6,12,13)
foodvarbl = (2*rep(1:13 %in% foodindex, 2)-1)
```

Next we set up indicator variables for the effect of bird species; this is the identity matrix of order 13 stacked on top of a replicate of itself.

```
birdvarbl = diag(rep(1,13))
birdvarbl = rbind(birdvarbl, birdvarbl)
```

Now we set up the 26 by 15 design matrix $\mathbf{Z}$ for the regression analysis. Its first column is all 1's in order to code the intercept or grand mean effect. Its second column contains 1's in the first 13 rows corresponding to the shellfish diet, and -1's in the remaining rows. The remaining 13 columns code the bird effects.

```
Zmat0          = matrix(0,26,15)
Zmat0[,1]      = 1
Zmat0[,2]      = foodvarbl
Zmat0[,3:15]   = birdvarbl
```

However, defining an effect for each bird in this way would make **Z** singular since the sum of these effects (columns 3:15) in each row is 1, and so is the intercept value. To correct for this, we need to force the sum of the bird effects within each diet group to add to zero. This requires two steps: we add two rows to the bottom of **Z** coding the sum of the bird effects for each diet group, and we add two corresponding functional observations to the 26 log count curves whose values are identically zero for all t.

```
Zmat           = rbind(Zmat0, matrix(0,2,15))
fishindex      = (1:13)[-foodindex]
Zmat[27,foodindex+2] = 1
Zmat[28,fishindex+2] = 1
birdextfd      = birdfd
birdextfd$coef =
    cbind(birdextfd$coefs, matrix(0,19,2))
```

Now we set up the arguments for fRegress. In these commands we insert each column in turn in matrix **Z** into the corresponding position in a list object xfdlist.

```
xfdlist = vector("list",15)
names(xfdlist) = c("const", "diet", birds)
for (j in 1:15) xfdlist[[j]] = Zmat[,j]
```

Now we define the corresponding list object betalist. We only want constant functions for the bird regression coefficient functions effects since there only the mean counts at the two sites available to estimate any bird's effect. However, for both the intercept and the food effect, we use a B-spline basis with knots at observation times. We determined the level of smoothing to be applied to the intercept and food regression functions by minimizing the cross-validated error sum of squares, as described in the next section, and the result was $\lambda = 10$.

```
betalist   = xfdlist
foodbasis  = create.bspline.basis(rng,21,4,yearCode)
foodfdPar  = fdPar(foodbasis, 2, 10)
betalist[[1]] = foodfdPar
betalist[[2]] = foodfdPar
conbasis   = create.constant.basis(rng)
for (j in 3:15) betalist[[j]] = fdPar(conbasis)
```

Next we fit the model using fRegress, which involves first defining lists for both the covariates (in this case all scalar) and a list of low-dimensional regression functions.

```
birdRegress = fRegress(birdextfd, xfdlist, betalist)
betaestlist = birdRegress$betaestlist
```

Figure 10.3 displays the regression functions for the intercept and food effects, along with 95% pointwise confidence intervals estimated by the methods described in Section 10.2.2. The trend in the intercept in the top panel models the mean trend over all species, and indicates a steady increase up to 1999 followed by some decline. The difference in the mean trends of the two food groups is shown in the bottom panel, and suggests a steady decline in the shellfish and mollusk eaters relative to the fish eaters starting in 1988. This is what we noticed in the log mean counts Figure 10.2.

Intercept

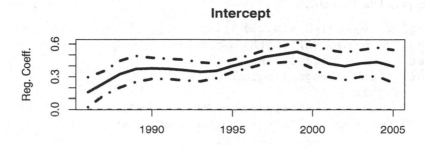

Food Effect

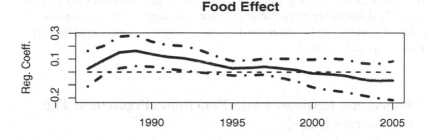

Fig. 10.3 The top panel shows the mean trend across all species, and the bottom panel shows the difference between being a shellfish/mollusk eater and a fish eater. The dashed lines indicate pointwise 95% confidence limits for these effects.

10.1.3 Choosing Smoothing Parameters

As for scalar response models, we would like to have a criterion for choosing any smoothing parameters that we use. Unfortunately, while ordinary cross-validation can be calculated for scalar response models without repeatedly re-estimating the model, this can no longer be done efficiently with functional response models. Here, we use the function fRegress.CV to compute the *cross-validated integrated squared error*:

$$\text{CVISE}(\lambda) = \sum_{i=1}^{N} \int \left(y_i(t) - \hat{y}_i^{(-i)}(t) \right)^2 dt$$

where $y^{(-i)}(t)$ is the predicted value for $y_i(t)$ when it is omitted from the estimation. In the following code, we search over a range of values for λ applied to both the intercept and the food effect:

```
loglam   = seq(-2,4,0.25)
SSE.CV1 = rep(NA,length(loglam))
betalisti = betaestlist
for (i in 1:length(loglam){
   for(j in 1:2)
      betalisti[[j]]$lambda = 10^loglam[i]
   CVi = fRegress.CV(birdRegress, xfdlist,
                     betalisti)
   SSE.CV1[i] = CVi$SSE.CV
}
```

This produces Figure 10.4, which indicates a unique minimum with λ approximately $\sqrt{10}$, although the discontinuities in the plot suggest that the cross-validated error sum of squares can be rather sensitive to non-smooth variation in the response functions as we defined them.

10.2 Functional Responses with Functional Predictors: The Concurrent Model

We can extend (10.1) to allow for functional covariates as follows:

$$y_i(t) = \beta_0(t) + \sum_{j=1}^{q-1} x_{ij}(t)\beta_j(t) + \varepsilon_i(t) \tag{10.4}$$

where $x_{ij}(t)$ may be a functional observation. Of course, x_{ij} may also be a scalar observation or a categorical indicator, in which case it can be simply interpreted as a function that is constant over time. Model (10.4) is called *concurrent* because it only relates the value of $y_i(t)$ to the value of $x_{ij}(t)$ at the same time points t. The intercept function $\beta_0(t)$ in effect multiplies a scalar covariate whose value is always one, and captures the variation in the response that does not depend on any of the covariate functions.

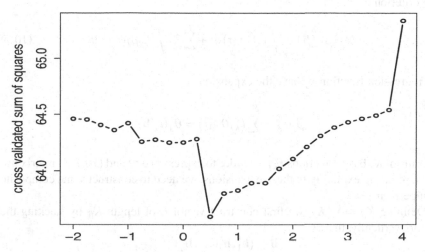

Fig. 10.4 The cross-validated integrated square errors for the bird data over a range of logarithms of λ.

10.2.1 Estimation for the Concurrent Model

As in ordinary regression, we must worry about redundancy or *multicollinearity* among the intercept and the functional (and scalar if present) covariates. Multicollinearity brings a host of problems, including imprecision in estimates due to rounding error, difficulty in discerning which covariates play an important role in predicting the dependent variable, and instability in regression coefficient estimates due to trade-offs between covariates in predicting variation in the dependent variable. When more than one functional covariate is involved, multicollinearity is often referred to as *concurvity*.

To better understand the multicollinearity problem, we look more closely at how the functional regression coefficients β_j are estimated by function fRegress by reducing the problem down to the solution of a set of linear equations. The coefficient matrix defining this linear system can then be analyzed to detect and diagnose problems with ill-conditioning and curvilinearity.

Let the N by q functional matrix $\mathbf{Z}$ contain these x_{ij} functions, and let the vector coefficient function β of length q contain each of the regression functions. The concurrent functional linear model in matrix notation is then

$$\mathbf{y}(t) = \mathbf{Z}(t)\beta(t) + \varepsilon(t) , \qquad (10.5)$$

where $\mathbf{y}$ is a functional vector of length N containing the response functions. Let

$$\mathbf{r}(t) = \mathbf{y}(t) - \mathbf{Z}(t)\beta(t) \qquad (10.6)$$

be the corresponding N-vector of residual functions. The weighted regularized fitting criterion is

$$\text{LMSSE}(\beta) = \int \mathbf{r}(t)'\mathbf{r}(t)\,dt + \sum_{j}^{p} \lambda_j \int [L_j\beta_j(t)]^2\,dt. \qquad (10.7)$$

Let regression function β_j have the expansion

$$\beta_j(t) = \sum_{k}^{K_j} b_{kj}\theta_{kj}(t) = \boldsymbol{\theta}_j(t)'\mathbf{b}_j$$

in terms of K_j basis functions θ_{kj}. In order to express (10.5) and (10.7) in matrix notation referring explicitly to these expansions, we need to construct some composite or supermatrices.

Defining $K_\beta = \sum_j^q K_j$, we first construct vector $\mathbf{b}$ of length K_β by stacking the vectors vertically, that is,

$$\mathbf{b} = (\mathbf{b}_1', \mathbf{b}_2', \ldots, \mathbf{b}_q')'.$$

Now assemble q by K_β matrix function $\Theta(t)$ as follows:

$$\Theta(t) = \begin{bmatrix} \boldsymbol{\theta}_1(t)' & 0 & \cdots & 0 \\ 0 & \boldsymbol{\theta}_2(t)' & \cdots & 0 \\ \vdots & \vdots & \cdots & \vdots \\ 0 & 0 & \cdots & \boldsymbol{\theta}_q(t)' \end{bmatrix}. \qquad (10.8)$$

Then $\beta(t) = \Theta(t)\mathbf{b}$. With this, (10.6) can be rewritten as follows:

$$\mathbf{r}(t) = \mathbf{y}(t) - \mathbf{Z}(t)\Theta(t)\mathbf{b}$$

Next let $\mathbf{R}(\lambda)$ be the block diagonal matrix with jth block as follows:

$$\lambda_j \int [L_j\boldsymbol{\theta}_j(t)]'[L_j\boldsymbol{\theta}_j(t)]\,dt.$$

Then (10.7) can be written as follows:

$$\text{LMSSE}(\beta) = \int [\mathbf{y}(t)'\mathbf{y}(t) - 2\mathbf{b}'\Theta(t)'\mathbf{Z}(t)'\mathbf{y}(t) + \mathbf{b}'\Theta(t)'\mathbf{Z}(t)'\mathbf{Z}(t)\Theta(t)\mathbf{b}]\,dt$$

$$+ \mathbf{b}'\mathbf{R}(\lambda)\mathbf{b}$$

If we differentiate this with respect to the coefficient vector $\mathbf{b}$ and set it to zero, we get the normal equations penalized least squares solution for the composite coefficient vector $\hat{\mathbf{b}}$:

$$[\int \Theta'(t)\mathbf{Z}'(t)\mathbf{Z}(t)\Theta(t)\,dt + \mathbf{R}(\lambda)]\hat{\mathbf{b}} = [\int \Theta'(t)\mathbf{Z}'(t)\mathbf{y}(t)\,dt]. \qquad (10.9)$$

This is a linear matrix equation defining the scalar coefficients in vector $\hat{\mathbf{b}}$, $\mathbf{A}\hat{\mathbf{b}} = \mathbf{d}$, where the normal equation matrix is

$$\mathbf{A} = \int \Theta'(t)\mathbf{Z}'(t)\mathbf{Z}(t)\Theta(t)\,dt + \mathbf{R}(\lambda)\,, \tag{10.10}$$

and the right-hand side vector of the system is

$$\mathbf{d} = \int \Theta'(t)\mathbf{Z}'(t)\mathbf{y}(t)\,dt\,. \tag{10.11}$$

These equations are all given in terms of integrals of basis functions with functional data objects. In some cases, it is possible to evaluate them explicitly, but we will otherwise revert to numerical integration. In practice, numerical integration is both feasible and accurate (with reasonable choices for basis sets, etc.).

Concurrent linear models make up an important subset of all possible linear functional response models, especially for examining dynamics (see Chapter 11). However, they can be particularly restrictive; we discuss the general class of linear functional response models in Section 10.3.

10.2.2 Confidence Intervals for Regression Functions

Confidence intervals for regression coefficients are produced by first estimating an error covariance and observing that our estimates are linear functions of our data. In doing this, we account for both the variation of the smoothed $\mathbf{y}(t)$ about their predicted values and the residuals of the original smoothing process. The residual for the ith observation of the jth curve is

$$r_{ij} = y_{ij} - \mathbf{Z}_j(t_i)\beta(t_i).$$

When the y_{ij} all occur at a grid of times, we can produce an estimate

$$\Sigma_e^* = \frac{1}{N}\mathbf{r}\mathbf{r}'. \tag{10.12}$$

Where $\mathbf{r}$ is the matrix of residuals.

In the Seabird data, the error variance is calculated from

```
yhatmat  = eval.fd(year, yhatfdobj)
rmatb    = logCounts2 - yhatmat
SigmaEb  = var(t(rmatb))
```

With this estimate of Σ_e^*, we must consider the smoothing done to take the observations of the response $\mathbf{y}$ onto the space spanned by the response basis functions $\phi(t)$. Let $\mathbf{C}$ denote the matrix of regression coefficients for this representation, so $\mathbf{y}(t) = \mathbf{C}\phi(t)$. Substituting this into (10.9), we get

$$\hat{\mathbf{b}} = \mathbf{A}^{-1} \Big[\int \Theta(t)' \mathbf{Z}(t)' \mathbf{C} \phi(t) dt \Big]$$

$$= \mathbf{A}^{-1} \Big[\int \phi(t)' \otimes (\Theta(t)' \mathbf{Z}(t)') dt \Big] \text{vec}(\mathbf{C})$$

$$= \texttt{c2bMap} \ \text{vec}(\mathbf{C}) \tag{10.13}$$

where $\otimes$ is used to represent the Kronecker product. The explicit use of a basis expansion for $y(t)$ allows the flexibility of modeling variation in $\mathbf{y}$ by itself or of including the original measurements of each response curve into the variance calculation.

We now require the matrix $\texttt{y2cMap}$ that is used compute the regression coefficient matrix $\mathbf{C}$ from the original observations, $\mathbf{y}$. This this can be obtained from functions like $\texttt{smooth.basis}$ or $\texttt{smooth.basisPar}$. The map ($\texttt{c2bMap}$ $\texttt{y2cMap}$) now maps original observations directly to $\hat{\mathbf{b}}$. Therefore:

$$\text{Var}\big[\hat{\mathbf{b}}\big] = \texttt{c2bMap} \ \texttt{y2cMap} \ \Sigma_e^* \ \texttt{y2cMap}' \ \texttt{c2bMap}'. \tag{10.14}$$

In the $\texttt{fda}$ package, these intervals are created using $\texttt{fRegress.stderr}$. This requires the result of a call to $\texttt{fRegress}$ along with the matrices $\texttt{y2cMap}$ and Σ_e^*. The standard errors for the regression coefficients used to create Figure 10.3 are computed using the following code.

```
y2cMap = birdList2$y2cMap
stderrList = fRegress.stderr(birdRegress, y2cMap,
                             SigmaEb)
betastderrlist = stderrList$betastderrlist
```

Finally we plot the results using the special purpose plotting function $\texttt{plotbeta}$.

When the original curves are not the result of smoothing data that have common observation times over curves, we can at least estimate confidence intervals based on the variation of the smoothed curves about the model predictions. To do this, we simply create pseudo data by evaluating the residual functions at a fine grid of points and calculating variance matrix from this. When we do this, the use of $\texttt{y2cMap}$ above is no longer valid. Instead we replace it with a projection matrix that takes the pseudo data to the coefficients $\mathbf{C}$. This is simply $[\Phi(\mathbf{t})'\Phi(\mathbf{t})]^{-1}$, but we will not pursue this here.

10.2.3 Knee Angle Predicted from Hip Angle

The gait data displayed in Figure 1.6 are measurements of angle at the hip and knee of 39 children as they walk through a single gait cycle. The cycle begins at the point where the child's heel under the leg being observed strikes the ground. For plotting simplicity we run time here over the interval [0,20], since there are 20 times at which the two angles are observed. This analysis is inspired by the question, "How much control does the hip angle have over the knee angle?"

Figure 10.5 plots the mean knee angle along with its angular velocity and acceleration, and Figure 10.6 plots knee angle acceleration against velocity. We can see three distinct phases in knee angle of roughly equal durations:

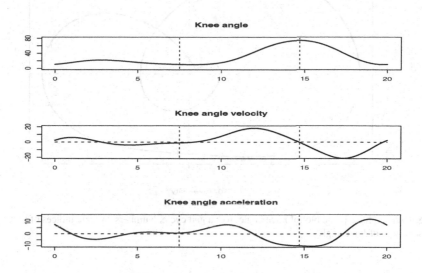

Fig. 10.5 Knee angle and its velocity and acceleration over a single gait cycle, which begins when the heel strikes the ground. The vertical dashed lines separate distinct phases in the cycle.

1. From time 0 to 7.5, the leg is bearing the weight of the child by itself, and the knee is close to being straight. This corresponds to the small loop in the cycle plot starting just before the marker "1" and up to the cusp.
2. From time 7.5 to time 14.7, the knee flexes in order to lift the foot off the ground, reaching a maximum mean angle of about 70 degrees.
3. From time 14.7 to time 20, the knee is extended to receive the load at the next heel-strike.

Together the second and third phases look like straightforward harmonic motion. A similar analysis of the hip motion reveals only a single harmonic phase. We wonder how the hip motion is coupled to knee motion.

Starting with functional data objects kneefd and hipfd for knee and hip angle, respectively, these commands execute a concurrent functional regression analysis where knee angle is fit by intercept and hip angle coefficient functions:

```
xfdlist       = list(rep(1,39), hipfd)
betafdPar     = fdPar(gaitbasis, harmaccelLfd)
betalist      = list(betafdPar,betafdPar)
fRegressList  = fRegress(kneefd, xfdlist, betalist)
kneehatfd     = fRegressList$yhatfd
```

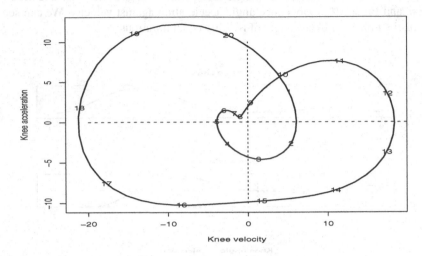

Fig. 10.6 A phase-plane plot of knee angle over a gait cycle. Numbers indicate indices of times of observation of knee angle.

```
betaestlist  = fRegressList$betaestlist
```

The intercept and hip regression coefficient functions are plotted as solid lines in Figure 10.7.

These commands compute the residual variance-covariance matrix estimate, which we leave as is rather than converting it to a diagonal matrix.

```
%kneemat     = eval.fd(gaittime, kneefd)
kneehatmat = eval.fd(gaittime, kneehatfd)
resmat.     = gait - kneehatmat
SigmaE      = cov(t(resmat.))
```

We also set up error sums of square functions for variation about both the model fit and mean knee angle. Then we compare the two via a squared multiple correlation function.

```
kneefinemat    = eval.fd(gaitfine, kneefd)
kneemeanvec    = eval.fd(gaitfine, kneemeanfd)
kneehatfinemat = eval.fd(gaitfine, kneehatfd)
resmat  = kneefinemat - kneehatfinemat
resmat0 = kneefinemat -
          kneemeanvec %*% matrix(1,1,ncurve)
SSE0 = apply((resmat0)^2, 1, sum)
SSE1 = apply(resmat^2, 1, sum)
Rsqr = (SSE0-SSE1)/SSE0
```

Intercept and Mean knee angle

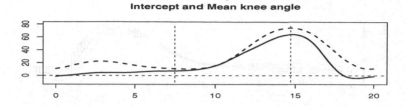

Hip coefficient and Squared multiple correlation

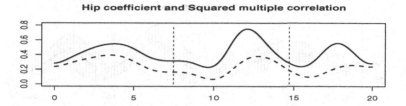

Fig. 10.7 The top panel shows as a solid line the intercept term in the prediction of knee angle from hip angle; the dashed line indicates the mean knee angle assuming no hip angle effect. The bottom panel shows as a solid line the functional regression coefficient multiplying hip angle in the functional concurrent linear model, and the dashed line shows the squared multiple correlation coefficient function associated with this model. Vertical dashed lines indicated boundaries between the three phases of the gait cycle.

The R^2 function is included in the second panel of Figure 10.7 as a dashed line. We see that it tracks pretty closely the variation in the hip regression coefficient.

The commands plot the intercept and hip regression coefficients with 95% confidence intervals, shown in Figure 10.8:

```
y2cMap = kneefdPar$y2cMap
fRegressList1 = fRegress(kneefd, xfdlist, betalist,
                         y2cMap, SigmaE)
fRegressList2 = fRegress.stderr(fRegressList1,
                         y2cMap, SigmaE)
betastderrlist = fRegressList2$betastderrlist
titlelist = list("Intercept", "Hip coefficient")
plotbeta(betaestlist, betastderrlist, gaitfine,
         titlelist)
```

We see that hip angle variation is coupled to knee angle variation in the middle of each of these three episodes, and the relation is especially strong during the middle flexing phase. It seems logical that a strongly flexed knee is associated with a sharper hip angle.

We can repeat these analyses to explore the relationship between knee and hip acceleration. This can be interesting because neural activation of these two muscle

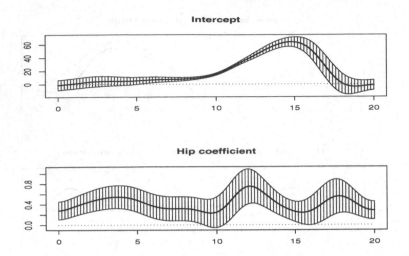

Fig. 10.8 The intercept and hip regression coefficient function for the gait cycle with 95% pointwise confidence intervals.

groups produces contraction, and contraction impacts acceleration directly by Newton's Second Law. Figure 10.9 shows the results of doing this. Now it is apparent that these two angles are coupled at a nearly evenly spaced series of seven points in the gait cycle. Given that a gait duration can be of the order of a second, it is striking to compare these plots with those of the handwriting data in Figure 8.6.

10.3 Beyond the Concurrent Model

The concurrent linear model only relates the value of a functional response to the current value of functional covariate(s). A more general version for a single functional covariate and an intercept is

$$y_i(t) = \beta_0(t) + \int_{\Omega_t} \beta_1(t,s)x_i(s)\mathrm{d}s + \varepsilon_i(t). \tag{10.15}$$

The bivariate regression coefficient function $\beta_1(s,t)$ defines the dependence of $y_i(t)$ on covariate $x_i(s)$ at each time t. In this case $x_i(s)$ need not be defined over the same range, or even the same continuum, as $y_i(t)$.

Set Ω_t in (10.15) contains the range of values of argument s over which x_i is considered to influence response y_i at time t, and the subscript t on this set indicates that this set can change from one value of t to another. For example, when both s and t are time, using $x_i(s)$ to predict $y_i(t)$ when $s > t$ may imply backwards causation.

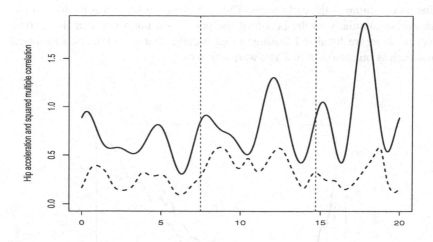

Fig. 10.9 The solid line shows the regression function multiplying hip angle acceleration in the prediction of knee angle acceleration, and the dashed line indicates the corresponding squared multiple coerrelation function.

To avoid this nonsense, we consider only values of x_i before time t. We may also add a restriction on how far back in time the influence of x_i on y_i can happen. This leads us to restrict the integral to

$$\Omega_t = \{s | t - \delta \leq s \leq t\},$$

where $\delta > 0$ specifies how much history is relevant to the prediction. Malfait and Ramsay (2003) described this as the *historical linear model*.

10.4 A Functional Linear Model for Swedish Mortality

We illustrate the estimation of (10.15) using Swedish life table data taken from census records in Sweden. The data are the number of deaths at each age for women born in each year from 1751 to 1914 and for ages 0 to 80. We will use the data up until 1884 in order to allow the extrapolation problem to be considered (see Section 10.10). Figure 10.10 displays the *log hazard rates* for four selected years. The log hazard rate is the natural logarithm of the ratio of the number of females who die at a specific age to the number of females alive with that age. These data were obtained from http://mortality.org. See also Chiou and Müller (2009) for another approach to modeling these data.

The hazard rate is greatest for infants and for the very elderly, and in most years attains its minimum in the early teens. The four curves indicate that the hazard rate decreases substantially as the health of the population improves over this period. However, there are localized features in each curve that reflect various historical events such as outbreaks of disease, war, and so on.

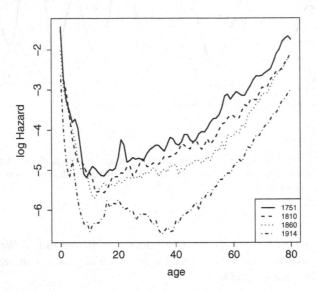

Fig. 10.10 Log hazard rates as a function of age for Swedish women born in the years 1751, 1810, 1860 and 1914. These data are derived from mortality tables at `http://mortality.org`.

Let $x_i(t)$ represent the log hazard rate at age t for birth year i. We propose the model

$$x_{i+1}(t) = \beta_0(t) + \int \beta_1(s,t)x_i(t)\mathrm{d}s + \varepsilon_i(t) \ . \qquad (10.16)$$

That is, for any year from 1752 to 1894, we model the log hazard function for that year using as the functional covariate the log hazard curve for the preceding year. Assume that the response curves have been smoothed and represented as functional data object `NextYear`, and that the covariate curves are in functional data object `ThisYear`.

The regression function β_1 has the basis function expansion

$$\beta_1(s,t) = \sum_{k=1}^{K_1} \sum_{\ell=1}^{K_2} b_{k\ell}\phi_k(s)\psi_\ell(t)$$

$$= \phi'(s)\mathbf{B}\psi(t), \qquad (10.17)$$

where the coefficients for the expansion are in the K_1 by K_2 matrix **B**. We therefore need to define two bases for β_1, as well as a basis for the intercept function β_0.

For a bivariate function such as $\beta_1(t,s)$ smoothness can be imposed by penalizing the s and t directions separately:

$$\text{PEN}_{\lambda_t,\lambda_s}(\beta_1(t,s)) = \lambda_1 \int [L_t\beta_1(t,s)]^2 \, ds \, dt + \lambda_2 \int [L_s\beta_1(t,s)]^2 \, ds \, dt , \qquad (10.18)$$

where linear differential operator L_s only involves derivatives with respect to s and L_t only involves derivatives with respect to t. We can also apply a penalty to the roughness of the intercept β_0.

The following code sets up a B-spline basis of order four with 23 basis functions. This is used to define functional parameter objects for β_0, $\beta_1(\cdot,t)$ and $\beta_1(s,\cdot)$. The second derivative is penalized in each case, but the smoothing parameter values vary as shown. The final statement assembles these three functional parameter objects into a list object to be supplied to function linmod as an argument.

```
betabasis = create.bspline.basis(c(0,80),23)
beta0Par  = fdPar(betabasis, 2, 1e-5)
beta1sPar = fdPar(betabasis, 2, 1e3)
beta1tPar = fdPar(betabasis, 2, 1e3)
betaList  = list(beta0Par, beta1sPar, beta1tPar)
```

Function linmod is invoked in R for these data by the command

```
linmodSmooth = linmod(NextYear, LastYear, betaList)
```

Figure 10.11 displays the estimated regression surface $\beta_1(s,t)$. The estimated intercept function β_0 ranged over values four orders of magnitude smaller than the response functions, and can therefore be considered to be essentially zero. The strong ridge one year off the diagonal, namely $\beta_1(s-1,s)$, indicates that mortality at any age is most strongly related to mortality at the previous year for that age less one. In other words, mortality is most strongly determined by age-specific factors like infectious diseases in infancy, accidents and violent death in early adulthood, and aging late in life. The height of the surface declines to near zero for large differences between s and t for this reason as well.

10.5 Permutation Tests of Functional Hypotheses

As was the case for scalar response models, we have so far focused on exploratory analyses. In the context of functional response models, it would again be useful to gauge the extent to which the estimated relationship can be distinguished from zero.

The type of questions that we are interested in are generalizations of common statistical tests and of common statistical models:

• Are two or more groups of functions statistically distinguishable?

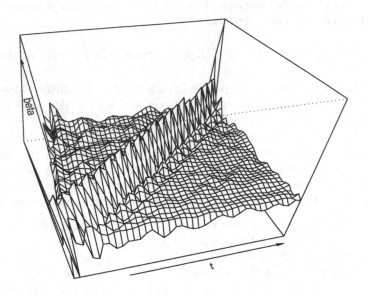

Fig. 10.11 The bivariate regression coefficient function $\beta_1(s,t)$ for the model (10.16) estimated from the 143 log hazard rate functions for the Swedish life table data. The ridge in $\beta_1(s,t)$ is one year off the diagonal.

- Are there statistically significant relationships among functional random variables?

Since functional data are inherently high dimensional, we again use permutation testing.

10.5.1 Functional t-Tests

Consider the Berkeley growth study data for both boys and girls in Figure 10.12. This plot suggests that boys generally become taller than girls. However, is this difference statistically significant? To evaluate this, we consider the absolute value of a t-statistic at each point:

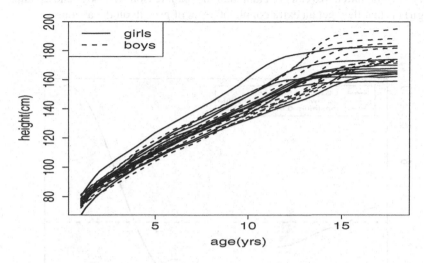

Fig. 10.12 The heights of the first ten boys and the first ten girls in the Berkeley growth study. We use a permutation test to evaluate whether the the growth curves for the two groups are different.

$$T(t) = \frac{|\bar{x}_1(t) - \bar{x}_2(t)|}{\sqrt{\frac{1}{n_1}\mathrm{Var}[x_1(t)] + \frac{1}{n_2}\mathrm{Var}[x_2(t)]}}. \tag{10.19}$$

This is plotted in the solid line in Figure 10.13. By itself, this provides a sense of the relative separation of the two groups of functions. However, a formal hypothesis test requires a value or statistic to test and a probability value indicating the result of the test. The test statistic that we use is the maximum value of the multivariate T-test, $T(t)$. To find a critical value of this statistic, we use a permutation test. We perform the following procedure:

1. Randomly shuffle the labels of the curves.
2. Recalculate the maximum of $T(t)$ with the new labels.

Repeating this many times allows a null distribution to be constructed. This provides a reference for evaluating the maximum value of the observed $T(t)$.

The following code executes a permutation test and generates the graphic in Figure 10.13. It uses a default value of 200 random shuffles, which is more than adequate for such a large difference as is shown, but might not suffice for more delicate effects.

```
tperm.fd(hgtmfd,hgtffd)
```

Here `hgtmfd` and `hgtffd` are functional data objects for the males and females in the study. It is apparent that there is little evidence for difference up to the age of around 12, about the middle of the female growth spurt, at which point the boys rapidly become taller. We can conclude that the main reason why boys end up taller than girls is that they get an extra couple of years of growth on the average.

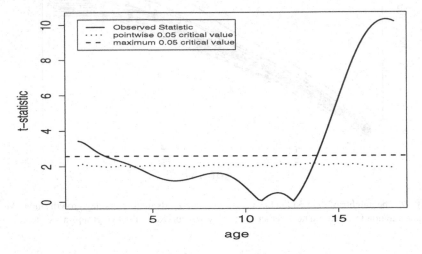

Fig. 10.13 A permutation test for the difference between girls and boys in the Berkeley growth study. The dashed line gives the permutation 0.05 critical value for the maximum of the *t*-statistic and the dotted the permutation critical value for the pointwise statistic.

10.5.2 Functional F-Tests

In the more general case of functional linear regression, the same approach can be applied. In this case, we define a functional version of the univariate *F*-statistic:

$$F(t) = \frac{\text{Var}[\hat{y}(t)]}{\frac{1}{n}\sum(y_i(t) - \hat{y}(t))^2} \qquad (10.20)$$

where $\hat{y}$ are the predicted values from a call to `fRegress`. Apart from a scale factor, this is the functional equivalent of the scalar *F*-statistic in multiple linear regression. It reduces to that for scalar-response models, as discussed in Section 9.5 above. As before, we reduce this to a single number by calculating $\max(F(t))$ and conducting a permutation test. In this case, we permute the response curves (or values), leaving the design unchanged. A test for no-effect of geographic region on temperature profile is conducted below. Figure 10.14 reports pointwise and maximal *F*-statistics and their corresponding permutation critical values for the temperature data.

```
F.res = Fperm.fd(temp36fd, regionList, betaList)
```

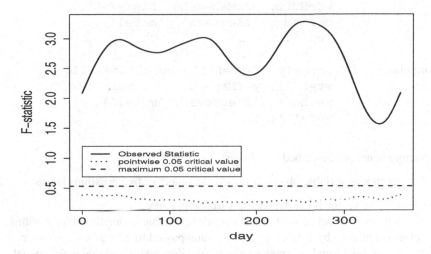

Fig. 10.14 A permutation test for a predictive relationship between geographic region and temperature profile for the Canadian weather data.

10.6 Details for R Functions fRegress, fRegress.CV and fRegress.stderr

10.6.1 Function fRegress

Because of its importance, we have set up a number of calling sequences for fRegress. These depend on the class of y, the first argument of the function. The first three cases concern the response being a vector of numbers, a functional data object or a functional parameter object. The last two allow the user to specify a model using a formula as in the R core function lm. The Matlab version can handle the first three cases, but not the last two:

```
numeric:    fRegress(y, xfdlist, betalist, wt=NULL,
                y2cMap=NULL, SigmaE=NULL, ...)

fd:     fRegress(y, xfdlist, betalist, wt=NULL,
                y2cMap=NULL, SigmaE=NULL, ...)
```

```
fdPar:          fRegress(y, xfdlist, betalist, wt=NULL,
                        y2cMap=NULL, SigmaE=NULL, ...)


character:      fRegress(y, data=NULL, betalist=NULL,
                        wt=NULL, y2cMap=NULL, SigmaE=NULL,
                        method=c('fRegress', 'model'),
                        sep='.', ...)


formula:        fRegress(y, data=NULL, betalist=NULL,
                        wt=NULL, y2cMap=NULL, SigmaE=NULL,
                        method=c('fRegress', 'model'),
                        sep='.', ...)
```

The arguments are described as follows:

y   The dependent variable object. It may be an object of five possible classes:

  scalar   A vector if the dependent variable is scalar.
  fd   A functional data object if the dependent variable is functional. A y of this
    class is replaced by fdPar(y, ...) and passed to fRegress.fdPar.
  fdPar   A functional parameter object if the dependent variable is functional,
    and if it is desired to smooth the prediction of the dependent variable.
  character or formula   A formula object or a character object
    that can be coerced into a formula providing a symbolic description of the
    model to be fitted satisfying the following rules: The left-hand side, formula
    y, must be either a numeric vector or a univariate object of class fd or fdPar.
    If the former, it is replaced by fdPar(y, ...).

    All objects named on the right-hand side must be either numeric or fd
    (functional data) or fdPar. The number of replications of fd or fdPar ob-
    ject(s) must match each other and the number of observations of numeric
    objects named, as well as the number of replications of the dependent variable
    object. The right-hand side of this formula is translated into xfdlist, then
    passed to another method for fitting (unless method = 'model'). Multi-
    variate independent variables are allowed in a formula and are split into
    univariate independent variables in the resulting xfdlist. Similarly, cate-
    gorical independent variables with k levels are translated into k−1 contrasts
    in xfdlist. Any smoothing information is passed to the corresponding com-
    ponent of betalist.

data   An optional list or data.frame containing names of objects identified
  in the formula or character y.
xfdlist   A list of length equal to the number of independent variables (including any intercept). Members of this list are the independent variables. They can
  be objects of either of these two classes:

  scalar   A numeric vector if the independent variable is scalar.
  fd   A (univariate) functional data object.

In either case, the object must have the same number of replications as the dependent variable object. That is, if it is a scalar, it must be of the same length as the dependent variable, and if it is functional, it must have the same number of replications as the dependent variable. (Only univariate independent variables are currently allowed in xfdlist.)

betalist   For the fd, fdPar, and numeric methods, betalist must be a list of length equal to length(xfdlist). Members of this list are functional parameter objects (class fdPar) defining the regression functions to be estimated. Even if a corresponding independent variable is scalar, its regression coefficient must be functional if the dependent variable is functional. (If the dependent variable is a scalar, the coefficients of scalar independent variables, including the intercept, must be constants, but the coefficients of functional independent variables must be functional.) Each of these functional parameter objects defines a single functional data object, that is, with only one replication.

For the formula and character methods, betalist can be either a list, as for the other methods, or NULL, in which case a list is created. If betalist is created, it will use the bases from the corresponding component of xfdlist if it is function or from the response variable. Smoothing information (arguments Lfdobj, lambda, estimate, and penmat of function fdPar) will come from the corresponding component of xfdlist if it is of class fdPar (or for scalar independent variables from the response variable if it is of class fdPar) or from optional . . . arguments if the reference variable is not of class fdPar.

wt   Weights for weighted least squares.

y2cMap   The matrix mapping from the vector of observed values to the coefficients for the dependent variable. This is output by function smooth.basis. If this is supplied, confidence limits are computed, otherwise not.

SigmaE   Estimate of the covariances among the residuals. This can only be estimated after a preliminary analysis with fRegress.

method   A character string matching either fRegress for functional regression estimation or mode to create the argument lists for functional regression estimation without running it.

sep   Separator for creating names for multiple variables for fRegress.fdPar or fRegress.numeric created from single variables on the right-hand side of the formula y. This happens with multidimensional fd objects as well as with categorical variables.

. . .   Optional arguments.

These functions return either a standard fRegress fit object or a model specification:

fRegress fit   A list of class fRegress with the following components:

y   The first argument in the call to fRegress (coerced to class fdPar).
xfdlist   The second argument in the call to fRegress.
betalist   The third argument in the call to fRegress.

betaestlist   A list of length equal to the number of independent variables and with members having the same functional parameter structure as the corresponding members of betalist. These are the estimated regression coefficient functions.

yhatfdobj   A functional parameter object (class fdPar) if the dependent variable is functional or a vector if the dependent variable is scalar. This is the set of predicted by the functional regression model for the dependent variable.

Cmatinv   A matrix containing the inverse of the coefficient matrix for the linear equations that define the solution to the regression problem. This matrix is required for function fRegress.stderr that estimates confidence regions for the regression coefficient function estimates.

wt   The vector of weights input or inferred.

df, OCV, gcv   Equivalent degrees of freedom for the fit, leave-one-out cross validation score for the model, and generalized cross validation score; only present if class(y) is numeric.

    If class(y) is either fd or fdPar, the fRegress object returned also includes 5 other components:

y2cMap   An input y2cMap.

SigmaE   An input SigmaE.

betastderrlist   An fd object estimating the standard errors of betaestlist.

bvar   The covariance matrix defined in (10.14).

c2bMap   A matrix converting the smoothing coefficients of the response variable into the regression coeffient, as defined in 10.13.

model specification The fRegress.formula and fRegress.character functions translate the formula into the argument list required by fRegress.fdPar or fRegress.numeric and then call the appropriate other fRegress function.

    Alternatively, to see how the formula is translated, use the alternative "model" value for the argument method. In that case, the function returns a list with the arguments otherwise passed to other fRegress methods (functions) plus the following additional components:

xfdlist0   A list of the objects named on the right hand side of formula. This will differ from xfdlist for any categorical or multivariate right-hand side object.

type   The type component of any fd object on the right-hand side of formula.

nbasis   A vector containing the nbasis components of variables named in formula having such components.

xVars   An integer vector with all the variable names on the right-hand side of formula containing the corresponding number of variables in xfdlist. This can exceed 1 for any multivariate object on the right-hand side of class either numeric or fd as well as any categorical variable.

10.6.2 Function *fRegress.CV*

`fRegress.CV` performs leave-one-out cross-validation for a model computed with `fRegress`. In the case of a scalar response, ordinary and generalized cross-validation scores can be computed analytically without having to leave each observation out one at a time. These scores are already returned by `fRegress`. For functional response models, we must calculate the cross-validated scores by brute force. This can take some time.

The function call is

```
fRegress.CV(y, xfdlist, betalist, CVobs,...)
```

The arguments are as follows:

y   The dependent variable object; either a vector, a functional data object or a functional parameter object.

xfdlist   A list whose members are functional parameter objects specifying functional independent variables. Some of these may also be vectors specifying scalar independent variables.

betalist   A list containing functional parameter objects specifying the regression functions and their level of smoothing.

CVobs   A vector giving the indexes of the observations to leave out, one at a time, in computing the cross-validation scores. This defaults to all observations, but may be used to leave out artificial zero observations, as in the functional ANOVA models described in this chapter.

...   Optional arguments not used by `fRegress.CV` but needed for superficial compatibability with `fRegress` methods.

The function returns a list object with components:

SSE.CV   The sum of squared errors, or integrated squared errors.

errfd.cv   Either a vector or a functional data object giving the cross-validated errors.

10.6.3 Function *fRegress.stderr*

The calling sequence is

```
fRegress.stderr(y, y2cMap, SigmaE, ...)
```

The arguments are as follows:

y   The named list of length six that is returned from a call to function `fRegress`.

y2cMap   A matrix that contains the linear transformation that takes the raw data values into the coefficients defining a smooth functional data object. Typically, this matrix is returned from a call to function `smooth.basis` that generates the dependent variable objects. If the dependent variable is scalar, this matrix is an identity matrix of order equal to the length of the vector.

SigmaE　　Either a matrix or a bivariate functional data object according to whether the dependent variable is scalar or functional, respectively. This object has a number of replications equal to the length of the dependent variable object. It contains an estimate of the variance-covariance matrix or function for the residuals.

...　　Optional arguments not used by fRegress.stderr but needed for superficial compatibability with fRegress methods.

The function returns a list object with these three components:

betastderrlist　　A list object of length equal to the number of independent variables. Each member contains a functional parameter object for the standard error of a regression function.

bvar　　A symmetric matrix containing sampling variances and covariances for the matrix of basis coefficients for the regression functions. These are stored columnwise in defining BVARIANCE.

c2bMap　　A matrix containing the mapping from response variable coefficients to coefficients for regression coefficients.

10.7 Details for Function plotbeta

The calling sequence is

```
plotbeta(betaestlist, betastderrlist, argvals=NULL,
         xlab="", ...)
```

The arguments are as follows:

betaestlist　　A list containing one or more functional parameter objects (class = fdPar) or functional data objects (class = fd).

betastderrlist　　A list containing functional data objects for the standard errors of the objects in betaestlist.

argvals　　A sequence of values at which to evaluate betaestlist and betastderrlist.

xlab　　x axis label.

...　　Additional plotting parameters passed to plot.

There is no return value.

10.8 Details for Function linmod

The calling sequence is

```
linmod(yfdobj, xfdobj, betaList)
```

The arguments are as follows:

yfdobj   A functional data object for the response or dependent variable functions.

xfdobj   A functional data object for the covariate or independent variable functions.

betaList   A list object containing three functional parameter objects. The first
is for the intercept term β_0 in (10.15), the second is for the bivariate regression
function β_1 in (10.15) as a function of the first argument s, and the third is for β_1
as a function of the second argument t.

The function returns a list of length three with components as follows:

beta0estfd   A functional data object for the estimated intercept.

beta1estbifd   A bivariate functional data object for the bivariate regression
function.

yhatfdobj   A functional data object for the predicted response function.

10.9 Details for Functions Fperm.fd and tperm.fd

10.9.1 Function Fperm.fd

This function can be called with exactly the same calling sequence as fRegress,
it has additional arguments which all have default values:

nperm Number of permutations to use in creating the null distribution.
argvals If yfdPar is a fd object, the points at which to evaluate the pointwise F-statistic.
q Critical upper-tail quantile of the null distribution to compare to the observed
F-statistic.
plotres Argument to plot a visual display of the null distribution displaying
the qth quantile and observed F-statistic.
... Additional plotting arguments that can be used with plot.

If yfdPar is a fd object, the maximal value of the pointwise F-statistic is calculated. The pointwise F-statistics are also returned. The default of setting q = 0.95
is, by now, fairly standard. The default nperm = 200 may be small, depending on
the amount of computing time available. If argvals is not specified and yfdPar
is a fd object, it defaults to 101 equally spaced points on the range of yfdPar.

If plotres = TRUE and yfdPar is a functional data object, a plot is produced giving the functional F-statistic along with 95th quantiles of the null distribution at each point and the 95th quantile of the null distribution of maximal F-values.
If yfdPar is scalar, a histogram is plotted with the 95th quantile marked along with
the observed statistic. The function returns a list with the following elements which
may be used to reconstruct the plot.

pval The observed p-value of the permutation test.

qval The qth quantile of the null distribution.
Fobs The observed maximal F-statistic.
Fnull A vector of length nperm giving the observed values of the permutation distribution.
Fvals The pointwise values of the observed F-statistic.
Fnullvals The pointwise values of of the permutation observations.
pvals.pts Pointwise p-values of the F-statistic.
qvals.pts Pointwise qth quantiles of the null distribution.
fRegressList The result of fRegress on the observed data.
argvals Argument values for evaluating the F-statistic if yfdPar is a functional data object.

10.9.2 Function tperm.fd

This function carries out a permutation t-test for the difference between two groups of functional data objects. Its arguments are

x1fd and x2fd Functional data objects giving the two groups of functional observations.
nperm The number of permutations to use in creating the null distribution.
q Critical upper-tail quantile of the null distribution to compare to the observed t-statistic.
argvals If yfdPar is a fd object, the points at which to evaluate the pointwise t-statistic.
plotres Argument to plot a visual display of the null distribution displaying the $1-q$th quantile and observed t-statistic.

If plotres=TRUE, a plot is given showing the functional t-statistic, along with the critical values of the permutation distribution at each point and the permutation critical value of the maximal t-statistic. It returns a list with the objects necessary to recreate the plot:

pval The observed p-value of the permutation test.
qval The qth quantile of the null distribution.
Tobs The observed maximal t-statistic.
Tnull A vector of length nperm giving the observed values of the permutation distribution.
Tvals The pointwise values of the observed t-statistic.
Tnullvals The pointwise values of of the permutation observations.
pvals.pts Pointwise p-values of the t-statistic.
qvals.pts Pointwise qth quantiles of the null distribution.
argvals Argument values for evaluating the F-statistic if yfdParis a functional data object.

10.10 Some Things to Try

The Swedish life table data consist of the log hazard rates (instantaneous risk of death) at ages 0 to 80 for Swedish women by birth year from 1751 to 1914. We want to develop a model for the way in which these have evolved over the years to 1894, and consider how well we can use this to forecast the hazard rate for women born in the year 1914.

1. Smooth the data appropriately. Explore these smooths – are there clearly evident features in how they change over time?
2. Create a functional linear model to predict the hazard curves from birth year for years 1751 to 1894. Choose smoothing parameters by cross-validation. Provide a plot of the error covariance. Plot the coefficient functions along with confidence intervals.
3. Examine the residuals from your model above. Are there any indications of lack of fit? If there are, construct an appropriately modified model. Plot the R-squared for both the linear model and the new one. Does there appear to be evidence for the effect of time on hazard curves?
4. Extrapolate your models to predict the hazard rate at 1914. How well does each do? Do they give better predictions than just the mean hazard curve?
5. Because the hazard curves are ordered in time, it is also possible to consider a *functional time series* model. Specifically, fit a model with the autoregressive structure:

$$y_{i+1}(t) = \beta_0(t) + \beta_1(t)y_i(t) + \varepsilon_i(t).$$

Report the results of your model. You should provide confidence intervals for $\beta_0(t)$ and $\beta_1(t)$. If you think the model will benefit from smoothing, do so. Does the model appear to fit well? Do you prefer this model or the model only using time as a predictor? Why?

10.11 More to Read

The concurrent linear model is closely related to the *varying coefficients model*. See Hastie and Tibshirani (1993), plus a large recent literature associated in Ramsay and Silverman (2005). A theoretical coverage of more general functional response models is given in Cuevas et al. (2002) as well as earlier papers by Cardot et al. (1999) and Ferraty and Vieu (2001). An elegant discussion of the ways in which the functional ANOVA can be treated is given in Brumback and Rice (1998) and associated discussion.

Chapter 11
Functional Models and Dynamics

This chapter brings us to the study of continuous time dynamics, where functional data analysis has, perhaps, its greatest utility by providing direct access to relationships between derivatives that could otherwise be studied only indirectly. Although dynamic systems are the subject of a large mathematical literature, they are relatively uncommon in statistics. We have therefore devoted the first section of this chapter to reviewing them and their properties. Then we address how "principal differential analysis (PDA)" can contribute to their study from an empirical perspective.

11.1 Introduction to Dynamics

Functional data offer access to estimated derivatives, which reflect *rates of change*. We have already seen the interpretative advantage of looking at velocity and acceleration in the Berkeley growth data. The field of *dynamics* is the study of systems that are characterized by relationships among derivatives. Newton's Second Law,

$$F = ma$$

which we can rewrite in functional data analysis terms as

$$D^2 x(t) = \frac{1}{m} F(t), \tag{11.1}$$

is probably the most famous dynamic model. It is, in fact, a concurrent functional linear model with a constant coefficient function where force F is the functional covariate predicting acceleration $D^2 x$. Dynamic models such as this are developed in the physical sciences from first principles and are often proposed as approximations to data derived from complex systems of all kinds.

J.O. Ramsay et al., *Functional Data Analysis with R and MATLAB*, Use R,
DOI: 10.1007/978-0-387-98185-7_11,
© Springer Science + Business Media, LLC 2009

11.1.1 An Example of First-Order Dynamics

Consider a straight-sided bucket of water with a leak at the bottom. Water will leak out from the hole at a rate proportional to the amount of pressure on the bottom of the bucket and the size of the hole (ignoring second-order effects like surface tension and flow turbulence). Since the pressure is proportional to the height $x(t)$ of the water in the bucket at time t, the flow rate $Dx(t)$ can be described as follows:

$$Dx(t) = -\beta x(t). \tag{11.2}$$

The negative sign is introduced here because water flowing *out* of the bucket reduces the height.

Equations with this structure have the solution

$$x(t) = Ce^{-\beta t}.$$

Since $C = x(0)$, it is called the *initial condition or state* of this system. Since in our example $\beta > 0$, the height of the water exhibits exponential decay.

If a hose adds water to the bucket at a rate $g(t)$, Equation (11.2) becomes

$$Dx(t) = -\beta x(t) + \alpha g(t). \tag{11.3}$$

The coefficient α is required to match the units of the two terms. The input function $g(t)$ is called a *forcing function*, changing the unforced behavior of the system.

Of course most buckets change their diameter with height, and there will be additional loss from evaporation, splashing and so forth. Effects such as these would require the coefficients to change with time:

$$Dx(t) = -\beta(t)x(t) + \alpha(t)g(t). \tag{11.4}$$

This is a concurrent functional linear model predicting the instantaneous rate of change $Dx(t)$ on the basis of two covariates: the *state* $x(t)$ and the external input or forcing $g(t)$.

Not all systems can be interpreted or developed so readily. Nonetheless, exploring the relationships between derivatives in a system can provide a useful guide to understanding its behavior. One of the useful aspects of first- and second-order linear dynamics is that explicit solutions can be given for systems like (11.3) that enable us to understand the instantaneous nature of the system's behavior. These are explored further in the next section.

Even fairly simple dynamic systems can produce highly complex behavior. This is one of the reasons they are such powerful mathematical tools. Analyzing such systems is an active area of research in applied mathematics with a large body of literature. However, for linear systems, it is fairly easy to write down explicit solutions.

When a forcing function is included the system, (11.3) leads to

$$x(t) = Ce^{-\beta t} + e^{-\beta t} \int_0^t \alpha e^{\beta \tau} g(\tau) d\tau.$$

To make this equation more interpretable, let us consider the situation where $g(t) = g$ is constant:

$$x(t) = Ce^{-\beta t} + \frac{\alpha g}{\beta}.$$

The height of water tends to a level $\alpha g / \beta$ that balances inflow of water with outflow leaving the bucket. Moreover, it tends to that level at an exponential rate. As a rule of thumb, the exponential term implies that $x(t)$ will move approximately two thirds of the distance to $\alpha g / \beta$ in $1/\beta$ time units.

11.1.2 Interpreting Second-Order Linear Dynamics

Of course, relationships between x and Dx may not capture all the important information about how a system evolves. *Linear second-order dynamics* are expressed as

$$D^2 x(t) = -\beta_0 x(t) - \beta_1 Dx(t) + \alpha g(t). \tag{11.5}$$

A good way to understand (11.5) is to think of a physical system described by Newton's Second Law: Each of the terms represents a different "force" on the system. The first term represents position-dependent forces like a spring for which the "stress" (force) is proportional to the "strain" (deformation). The second term is proportional to the speed at which the system moves, and can be thought of in terms of friction or viscosity, especially when β_1 is positive. As before, $g(t)$ again represents external inputs into a system that modify its behavior, like Newton's Second Law, expression (11.1) above.

Let us first suppose that $\beta_1 = \alpha = 0$ so that

$$D^2 x(t) = -\beta_0 x(t).$$

If $\beta_0 \geq 0$, the solutions are of the form

$$x(t) = c_1 \sin(\sqrt{\beta_0} t) + c_2 \cos(\sqrt{\beta_0} t),$$

which are periodic with a period $1/\sqrt{\beta_0}$.

More generally, if $\beta_1 \neq 0$, we examine the *discriminant*

$$d = \beta_1^2 / 4 - \beta_0.$$

Direct differentiation shows that solutions are given by linear combinations of exponential functions

$$x(t) = c_1 \exp(\gamma_1 t) + c_2 \exp(\gamma_2 t)$$

with

$$\gamma_1 = \frac{-\beta_1}{2} + \sqrt{d}, \ \gamma_2 = \frac{-\beta_1}{2} - \sqrt{d}.$$

These solutions will decay exponentially if $\gamma_1 < 0$ (since $\gamma_2 <= \gamma_1$). If $d < 0$, γ_1 and γ_2 are complex conjugates, and

$$x(t) = \exp(-\beta_1 t/2)[d_1 \sin(t\sqrt{-d}) + d_2 \cos(t\sqrt{-d})]. \qquad (11.6)$$

This yields oscillations that increase or decrease according to the sign of β_1. Moreover, the oscillations have period $2\pi/\sqrt{-d}$.

Using these observations, we can divide the (β_0, β_1) space into regions of different qualitative behavior: oscillatory and nonoscillatory, exponential growth and exponential decay. This division is depicted in Figure 11.1.

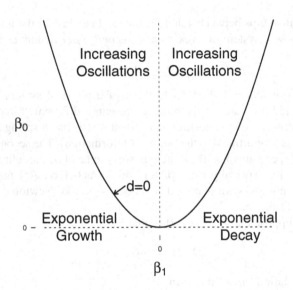

Fig. 11.1 A diagram of the various dynamic regimes for a second-order differential equation for different values of β_0 and β_1.

For many purposes, we may want to generalize expression (11.5) further to consider time-varying coefficients:

$$D^2 x(t) = -\beta_1(t)Dx(t) - \beta_0(t)x(t) \qquad (11.7)$$

in terms of the instantaneous values of the functional discriminant

$$d(t) = \beta_1(t)^2/4 - \beta_0(t).$$

These diagnostics should be understood with some caution: If the coefficient functions in (11.7) vary rapidly, their instantaneous changes may not translate into substantive changes in the overall behavior of $x(t)$. For example, if $d(t)$ becomes negative only briefly, there may not be enough time for any meaningful oscillation to occur. Rapidly changing coefficient functions or strong relationships between the coefficient functions of $x(t)$ and its derivatives may provide a good indication that a more complex, nonlinear system could be considered.

We draw from all of this that a relatively simple dynamic equation like (11.5) can define a wide variety of behaviors. We also see the important issue of system *stability*. There are multiple definitions of stability, but in general if the coefficient of velocity, $\beta_1(t)$, is positive, the system will be stable, exhibiting something like exponential decay possibly with a damped oscillation; if $\beta_1(t)$ is negative, the system will exhibit something like exponential growth or a similarly growing oscillation.

11.1.3 Higher-Dimensional Linear Systems

Dynamic models can include more than one state variable. The second-order system discussed in Section 11.1.2 can be cast as a first-order system with a vector state, with components representing the location and velocity. In this context it is less easy to produce analytic expressions with which to analyze the stability properties of a system. However, the rules are not very different. A multidimensional linear system involving a k-dimensional state can be written as

$$D\mathbf{x}(t) = -\mathbf{B}(t)\mathbf{x}(t), \tag{11.8}$$

where $\mathbf{B}(t)$ is now a $k \times k$ matrix. It is clear that for this system, $\mathbf{x}(t) \equiv 0$ results in a stable solution.

We can specialize this system to constant coefficients and add a forcing function to get the following:

$$D\mathbf{x}(t) = -\mathbf{B}\mathbf{x}(t) + \mathbf{u}(t).$$

If $\mathbf{u}$ is constant, this system has a *fixed point* at $\mathbf{x} = -\mathbf{B}^{-1}\mathbf{u}$ at which the solution does not change. We can understand the stability of this solution in terms of the eigenvalues $d_1, \ldots, d_k$ of $-\mathbf{B}$. Letting

$$\xi_j(t) = e^{d_j t}, \ j = 1, \ldots, k,$$

the solution to (11.8) is given in terms of linear combinations of the $\xi_j(t)$. For a general matrix $\mathbf{B}$, some of the eigenvalues may be complex. For real-valued matrices $\mathbf{B}$, any complex eigenvalue will be paired with its complex conjugate. Moreover, the imaginary parts describe the oscillations that we observed for the second-order system, with the period of oscillation being 2π over the (positive) imaginary part of each complex conjugate pair. Moreover, any eigenvalue with a positive real part will explode exponentially; a complex conjugate pair of eigenvalues with a positive real part will exhibit an exponentially increasing oscillation. A forcing term may shift the behavior but will not change the stability properties unless the forcing term is a function of the state vector in a way that in essence modifies the state transition matrix $\mathbf{B}$.

How are we to deal with higher-order multivariate dynamics? In Section 11.4 we use a model of the form

$$D^2\mathbf{x}(t) = -\mathbf{B}_0(t)\mathbf{x}(t) - \mathbf{B}_1(t)D\mathbf{x}(t) + \mathbf{u}(t).$$

In order to examine the stability of this system, we expand it by creating a new variable $\mathbf{y}(t) = D\mathbf{x}(t)$. This system can be written down as

$$\begin{pmatrix} D\mathbf{y}(t) \\ D\mathbf{x}(t) \end{pmatrix} = \begin{pmatrix} -\mathbf{B}_1(t) & -\mathbf{B}_0(t) \\ \mathbf{I} & 0 \end{pmatrix} \begin{pmatrix} \mathbf{y}(t) \\ \mathbf{x}(t) \end{pmatrix} + \begin{pmatrix} \mathbf{u}(t) \\ 0 \end{pmatrix}.$$

That is, we treat $D\mathbf{x}(t)$ as extra dynamic variables to provide a first-order $2k \times 2k$ system. The same analysis may be made of the eigenvalues of the expanded matrix above. As for one-dimensional systems, we can only interpret the local behavior of a system if the parameters in the system change at a much slower rate than the system itself.

11.2 Principal Differential Analysis for Linear Dynamics

We have seen how linear models describing relationships between derivatives results in a system whose behavior can be qualitatively characterized. We would now like to use this theory to characterize the behavior of a system from which we have data.

How can we fit linear dynamic models to functional data? One approach is to solve a differential equation like (11.3) for some value of the parameters and fit this to observed data by least squares. This procedure is computationally expensive, however, and such models rarely fit observed data well since they do not account for unobserved external influences on a system.

Instead, we use the fact that functional data analysis already gives us derivative information. Given repeated measurements of the same process, we can model

$$Dx_i(t) = -\beta(t)x_i(t) + \alpha(t)u_i(t) + \varepsilon_i(t), \tag{11.9}$$

where the $\varepsilon_i(t)$ are error terms to allow for variation between different curves. This expression represents a functional linear regression and could be fit with fRegress.

However, we can view the model in a different light: when $u_i(t) \equiv 0$ functional linear regression estimates $\beta(t)$ to minimize

$$\text{PDASSE}(\beta) = \sum_{i=1}^{N} \int \left[Dx_i(t) + \beta(t)x_i(t) \right]^2 dt = \sum_{i=1}^{N} \int \left[L_\beta x_i(t) \right]^2 dt. \tag{11.10}$$

That is, the model looks for a linear differential operator to represent covariation between x and Dx. This method has been labeled *principal differential analysis* (PDA) because of its similarity to principal components analysis:

- Functional PCA looked for linear operators defined by $\beta(t)$ to explain variation between curves.
- PDA looks for linear operators to explain variation between derivatives but within curves.

Naturally, we can extend the same ideas to multivariate functions and to higher derivatives; these are all accommodated in the `fda` package.

When we also wish to consider inputs into a dynamic system, the PDA objective criterion is the difference between the effective input and the linear differential operator:

$$\text{PDASSE}_u(\beta) = \sum_{i=1}^N \int \left[L_\beta x_i(t) - \alpha(t)u(t) \right]^2 dt. \qquad (11.11)$$

Both β and α here are functional objects to be estimated. This creates an *input-output* system which responds to changes in $u(t)$. Our examples below do not use forcing functions, but we provide a description of how to incorporate them into the code.

11.3 Principal Differential Analysis of the Lip Data

We illustrate the use of PDA with data on the movement of lips during speech production. Figure 11.2 presents the position of the lower lip when saying the word "Bob" 20 times. As is clear from the data, there are distinct opening and shutting phases of the mouth surrounding a fairly linear trend that corresponds to the vocalization of the vowel. Muscle tissue behaves in many ways like a spring. This observations suggests that we consider fitting a second-order equation to these data.

The function `pda.fd` is the basic tool for this analysis. In a break from our naming conventions, the equivalent Matlab function is `pdacell`. The arguments to this function are similar to `fRegress`. We need to give it the functional data object to be analyzed along with a list of functional parameter objects containing bases and penalties for the β and α coefficient functions. The following code attempts to derive a second-order homogeneous differential equation like expression (11.7) for `lipfd` obtained from smoothing the lip data with no smoothing in the coefficients $\beta_0(t)$ and $\beta_1(t)$:

```
lipfd    = smooth.basisPar(liptime, lip, 6,
             Lfdobj=int2Lfd(4), lambda=1e-12)$fd
names(lipfd$fdnames) = c("time(seconds)",
           "replications", "mm")
lipbasis = lipfd$basis
bwtlist  = list(fdPar(lipbasis,2,0),
```

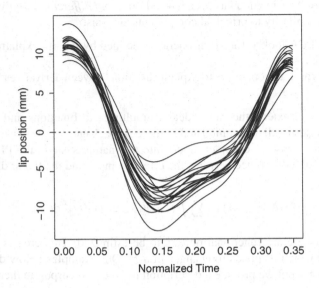

Fig. 11.2 The lip data. These give the position of the lower lip relative to the upper during 20 enunciations of the word "Bob" by the same subject.

$$\texttt{fdPar(lipbasis,2,0))}$$
```
pdaList   = pda.fd(lipfd,bwtlist)
```

The definition of `pda.fd` provides for arguments `awtlist` and `ufdlist`, whose absence here indicates that the forcing function $\alpha g(t)$ in (11.5) is zero.

We now need to analyze the result. The function

```
plot.pda.fd(pdaList,whichdim=3)
```

will plot the first two panels in Figure 11.3. In higher dimensional systems these coefficient functions can be grouped by dimension, equation, or observation. For the third panel we have plotted the discriminant function:

```
dfd         = (0.25*pdaList$bwtlist[[2]]$fd^2
              - pdaList$bwtlist[[1]]$fd )
dfd$fdnames= list('time','rep','discriminant')
```

From this we see that there is an initial explosive motion as the lips, previously sealed, are opened. This is followed by a period where the motion of the lips is largely oscillatory with a period of about 30-40 ms. This corresponds approximately to the spring constant of flaccid muscle tissue. During the "o" phase of the word, there is a period of damped behavior when the lips are kept open in order to enunciate the vowel.

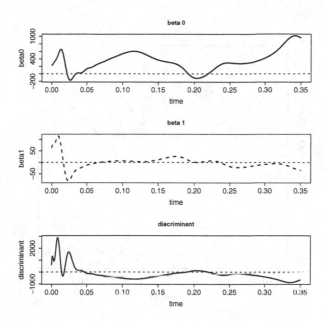

Fig. 11.3 Results of performing principal differential analysis on the lip data. Top two panels represent the estimated $\beta_0(t)$ and $\beta_1(t)$ functional coefficients. The bottom panel shows the discriminant function. This reveals in initial explosive motion as the lips part followed by oscillatory motion, modulated for the production of the "o".

We can also overlay our β coefficients on the bifurcation diagram in Figure 11.1. The following code produces Figure 11.4:

```
pda.overlay(pdaList)
```

This tells much the same story. The initial impulse corresponds to explosive growth, followed by largely stable oscillatory motion.

11.4 PDA of the Handwriting Data

PDA could be effectively carried out using fRegress. However pda.fd also works for for multivariate functional observations, which cannot be handled by the current version of fRegress. Here we examine the handwriting data in Figure 1.8 where PDA provides informative results. As for the lip data, since this is a physical system, we model the second derivative of the data.

For multidimensional systems, a PDA will have three levels of weight functions. Each function is indexed according to the equation in which it appears (i), the variable it multiplies (j), and the derivative of that variable (k):

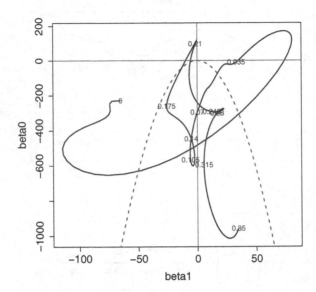

Fig. 11.4 An overlay of the PDA parameters estimated for the lip data on a bifurcation diagram of a second-order linear system. Points are marked as time points from 0 to 0.35 seconds. The initial explosive growth is followed by a period close to undamped oscillations.

$$Lx_{qi}(t) = D^m x_{qi}(t) + \sum_{k=0}^{m-1} \sum_{j=1}^{d} \beta_{ijk}(t) D^k x_{qj}(t). \tag{11.12}$$

the subscript q is used here to represent repeated observations of a multivariate functional process.

In order to account for these levels of functions, for multidimensional systems pda.fd uses lists in R and cell arrays in Matlab for both functional data and functional parameter objects. To begin with, we take x to be given by a list of functional data objects, one for each dimension. This allows the various dimensions of x to be defined with respect to different bases. In the handwriting data example we split the fdafd object into each of its dimensions:

```
xfdlist = list(fdafd[,1],fdafd[,2])
```

If we want to construct a second-order analysis, we need a three-dimensional array of functional parameter objects. In Matlab, this is simply a three-dimensional cell array, the dimensions being in the order (i, j, k) in (11.12). In R, we achieve the same result by nesting lists within lists: Once we have created the bwtlist object, we should have that bwtlist[[i]][[j]][[k]] contains the functional parameter object needed to define $\beta_{ijk}(t)$. The following code sets up the analysis of the handwriting data:

```
pdaPar = fdPar(fdabasis,2,1)
pdaParlist = list(pdaPar, pdaPar)
bwtlist = list( list(pdaParlist,pdaParlist),
                list(pdaParlist,pdaParlist) )
pdaList = pda.fd(xfdlist, bwtlist)
```

For higher-dimensional systems, the analysis presented in Figure 11.4 is no longer feasible. Instead, we consider the pointwise eigenvalues of the system that we described in Section 11.1.3. These can be plotted as functional quantities. Nonzero eigenvalues sometimes come in conjugate pairs. Therefore, a plot of the imaginary part of the eigenvalues may be symmetrical. With nonzero imaginary parts, the system oscillates. When the real part of any eigenvalue is positive, the system experiences exponential growth or a growing oscillation. Otherwise it is stable or decaying.

The function eigen.pda(pdaList) takes the result of pda.fd and produces the stability analysis given in Figure 11.5. As can be see, there is a strong and stable periodic solution in the data, with the real parts of the eigenvalues staying close to zero, indicating that the writing is dominated by ellipsoidal motion.

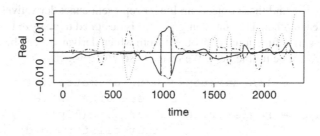

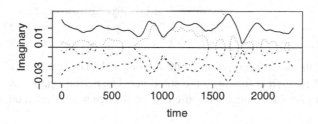

Fig. 11.5 Stability analysis of a principal differential analysis (PDA) of the handwriting data. The nearly constant imaginary eigenvalue indicates a constant cycle modulated locally.

11.5 Registration and PDA

How do we reconcile registration and the analysis of dynamics? These appear to be competing demands. The appearance of features in data such as the pubertal growth spurt makes an *a priori* case for registration: the dynamics of growth are likely to be markedly different during puberty than at other times. We therefore would like to align individuals so that puberty occurs at the same time and the dynamics should therefore be comparable across individuals. However, it is fairly easy to see that registration can have a strong effect on the derivatives of functional data:

$$Dx(h(t)) = D[h](t)D[x](h(t))$$

Here, $D[x](h(t))$ indicates that this is the derivative of x with respect to its argument rather than with respect to t. This will be stronger for second derivatives. We may thus lose the comparison that registration was designed to achieve.

One way around this is to register derivatives individually. That is, we first take derivatives, register each of them individually with the same registration curve, and then conduct a regression analysis. The `register.newfd` function is designed to do this. The following code carries this out for the lip data. We first perform landmark registration, then register each of the first and second derivatives with the resulting registration function. Doing this creates a separate functional data object for each derivative, and these objects no longer represent exact derivatives and anti-derivatives of each other. The function `pda.fd` is designed to analyze the dynamic properties of a single functional data object from which derivatives can be extracted. It will therefore not be usable here. We instead use `fRegress`:

```
lipreglist = landmarkreg(lipfd, as.matrix(lipmarks),
                         lipmeanmarks, WfdPar)
Dlipregfd   = register.newfd(deriv.fd(lipfd,1),
                         lipreglist$warpfd)
D2lipregfd  = register.newfd(deriv.fd(lipfd,2),
                         lipreglist$warpfd)
xfdlist     = list(-Dlipregfd,-lipreglist$regfd)
lipregpda   = fRegress( D2lipregfd, xfdlist, bwtlist)
```

The results of this have been plotted with the original PDA results in Figure 11.6: The registered coefficient functions are smoother. The period near the end in which $\beta_0(t)$ is close to zero also suggests a pure frictional force when the mouth is closing.

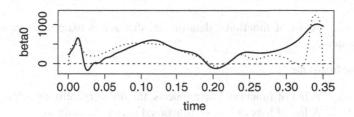

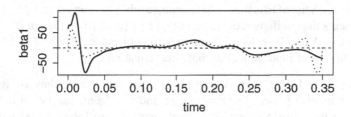

Fig. 11.6 A comparison of PDA results for the unregistered lip data (solid) and the lip data with each derivative first calculated and then registered with the same warping function (dashed).

11.6 Details for pda.fd, eigen.pda, pda.overlay and register.newfd

11.6.1 Function pda.fd

The pda.fd function is like fRegress except that since both the response and covariates are given by derivatives of the same function, only one functional data object needs to be specified. It also handles multivariate functional data, but insists that each dimension be given in a separate element of a list. This allows each dimension to be defined with respect to different bases. The standard call is

    pda.fd(xfdlist, bwtlist,awtlist, ufdlist,nfine)

We divide the description of the arguments into two cases:

$x(t)$ is univariate:

xfdlist   A functional data object defining $x(t)$.
bwtlist   A list of functional parameter objects, the jth element of each should specify the basis and smoothing penalty for $\beta_j(t)$.

awtlist   A list of functional parameter objects defining the coefficient functions for the inputs, this should be the same length as ufdlist and may be NULL.

ufdlist   A list of functional data objects that act as external influences on the system.

$x(t)$ is multivariate:

xfdlist   A list of functional data objects, the ith entry defining $x_i(t)$.

bwtlist   A list of lists of lists of functional parameter objects.
    bwtlist[[i]][[j]][[k]] should define the basis and smoothing penalty for $\beta_{ijk}(t)$.

awtlist   A list of lists functional parameter objects. awtlist[[i]][[j]] represents the coefficient of ufdlist[[i]][[j]] in equation i.

ufdlist   A two-level list of functional data objects, ufdlist[[i]] represents the list of input functions that affect equation i.

Both awtlist and ufdlist default to NULL, in which case they are ignored. Individual elements of bwtlist, awtlist and ufdlist can be set to NULL, in which case the corresponding coefficient functions are forced to be zero. The nfine component gives the number of evaluation points at which to perform a linear regression. It defaults to 501.

The function returns:

bwtlist   A list array of the same dimensions as the corresponding argument, containing the estimated or fixed weight functions defining the system of linear differential equations.

resfdlist   A list of length equal to the number of variables or equations. Each member is a functional data object giving the residual functions or forcing functions defined as the left side of the equation (the derivative of order m of a variable) minus the linear fit on the right side. The number of replicates for each residual functional data object is the same as that for the variables.

awtlist   A list of the same dimensions as the corresponding argument. Each member is an estimated or fixed weighting function for a forcing function.

11.6.2 Function eigen.pda

This function calculates the pointwise eigenvalues of the system and produces a plot of the same format as Figure 11.5. If awtlist is present, the fixed point of the system is also calculated at each time and plotted. Its arguments are

pdaList   A list object returned by pda.fd.
plotresult   Should the result be plotted? Default is TRUE.
npts   Number of points to use for plotting.

... Other arguments for `plot`.

In addition to producing the plot the function returns a list with elements

`argvals` The evaluation points of the coefficient functions.
`eigvals` The corresponding eigenvalues of the system at each point.
`limvals` The pointwise fixed-point of the system.

11.6.3 Function `pda.overlay`

For a second-order univariate principal differential analysis, this function plots $\beta_0(t)$ against $\beta_1(t)$ and overlays a bifurcation diagram. It requires

`pdaList` A list object returned by `pda.fd`.
`nfine` Number of points to use for plotting.
`ncoarse` Number of points use as time markers along the plot.

11.6.4 Function `register.newfd`

This function will register a given functional data object with a specified warping function. It requires

`yfd` A multivariate functional data object defining the functions to be registered with `Wfd`.
`Wfd` A functional data object defining the registration functions to be used to register `yfd`. This can be the result of either `landmarkreg` or `register.fd`.
`type` Indicates the type of registration function.

  `direct` Assumes `Wfd` is a direct definition of the registration functions. This is produced by `landmarkreg`.
  `monotone` Assumes that `Wfd` defines a monotone functional data objected, up to shifting and scaling to make end points agree. This is produced by `register.fd`.
  `periodic` Does shift registration for periodic functions. This is output from `register.fd` if `periodic=TRUE`.

It outputs a functional data object containing the registered curves.

11.7 Some Things to Try

1. Instead of the time-varying principal differential analysis given for the handwriting data, try a constant-coefficient principal differential analysis, but include a

constant forcing term. Does your interpretation differ markedly? What does the fixed point of the system tell you?

2. Try a PDA of the Chinese handwriting data. Do the dynamics of this system appear to be very different from the cursive script?

3. PDA can be performed on a single time series as well, but we have to borrow strength across times instead of across replicates. One easy way to do this is to insist that all the $\beta_{ijk}(t)$ be constant. Try this with data on the incidence of melanoma over a 30-year period. These data are available in the `melanoma` object.

   a. Smooth the data, choosing the optimal λ by `gcv` and plot both data and the smooth. We observe that there are two distinct dynamics: a linear trend and a cycle with a period of about 10 years.

   b. These observations suggest that

   $$D^4x(t) + \alpha D^2(x) \approx 0.$$

   We would like to get a handle on α. To do this, conduct a PDA using your estimated smooth and representing α as a constant functional data object.

   c. The `pda.fd` function produces an `Lfd` object. Try resmoothing the data using this object to define a penalty. How does the optimal value of λ change? How do the degrees of freedom change?

11.8 More to Read

The study of dynamics has a long history in applied mathematics. Borrelli and Coleman (2004) provide a good introductory overview. While linear differential equations with constant coefficients are relatively easily studied, nonlinear systems are harder to analyze. Generalizing (11.8) a system is described by a vector of states, and its evolution is given in terms of

$$D\mathbf{x} = \mathbf{f}(\mathbf{x}, \mathbf{u}, \theta), \tag{11.13}$$

where $\mathbf{f}$ is a vector-valued nonlinear function. Unlike (11.8), however, it is not usually possible to write down solutions to (11.13) analytically. Instead, we must rely on numerical methods to approximate them. Despite these challenges, nonlinear dynamic systems have proved enormously versatile in producing different forms of qualitative behavior that mimic real-world processes, from bursts of neural firing through epidemic processes and chemical reactions. We can examine the behavior of these systems by extending the analysis of the stability of linear systems that we described above and examining how the stability of fixed points and cycles changes with elements of the parameter vector θ. This is a large field and the reader is directed to the relevant literature such as Kuznetsov (2004) to learn more.

Despite the usefulness of such models, there is relatively little literature on assessing their agreement with data or on estimating and performing inference for θ. This is partly due to the numerical difficulties involved in finding solutions to (11.13) and partly due to the idealization involved in assuming that a system evolves deterministically. One way of reducing the numerical challenges in fitting these data is a nonlinear version of PDA; if all the components of $\mathbf{x}$ are measured, we can smooth the data to create $\hat{\mathbf{x}}$ and estimate θ to minimize

$$\mathrm{SISE}(\theta) = \int \left(D\hat{\mathbf{x}}(t) - \mathbf{f}(\mathbf{x}(t), \mathbf{u}(t), \theta) \right)^2 dt.$$

The idea has been rediscovered numerous times (see Bellman and Roth, 1971; Varah, 1982; Pascual and Ellner, 2000; Chen and Wu, 2008). The statistical properties of the resulting estimates have recently been examined (Brunel, 2008). This technique can only be used, however, when there are enough data to smooth each component of $\mathbf{x}$. More recent work has focused on using (11.13) as a smoothing penalty and iteratively refining θ to match the data (Ramsay et al., 2007). The use of functional data analysis in statistical inference for nonlinear systems remains an important research area.

Symbol Table

Numbers in parentheses refer to chapters where the symbol is used as indicated.

$b, \mathbf{b}$	regression coefficient function(s) estimates
$c, \mathbf{c}$	basis expansion coefficient(s)
d	discriminant of a second-order system; eigenvalue for a first-order system
g	forcing function
$h, \mathbf{h}$	warping function(s)
i, j, k, ℓ	indices
I, J, K, m, n, N	dimensions of vectors or matrices
$s, \mathbf{s}$	value(s) on the domain of a function
$t, \mathbf{t}$	value(s) on the domain of a function
w, W	log derivative of monotone or warping function
$x, \mathbf{x}$	functional data observation(s)
$y, \mathbf{y}$	functional data observation
$z, \mathbf{z}$	covariate scalar or functional data observation(s)
α	rate constant in an exponent (3); an intercept (9); forcing function (11)
$\beta, \boldsymbol{\beta}$	regression coefficient function (scalar or vector)
γ	rate constant in an exponent
δ	time shift (8, 10); statistical technique (10)
ε	error or residual
θ	latent ability value (1); parameter (11)
λ	smoothing parameter value
μ	mean function (9,10,1); eigenvalue (7)
ν	eigenvalue (7)
ξ	weight function (6); exponential basis function (11)
η	weight function (7)
π	trigonometric constant
ρ	correlation (4, 6); probe functional (6, 7)
σ, Σ	standard deviation, variance, covariance

ϕ, φ	basis function
ψ, ζ	basis function
Θ	matrix of basis function values
Φ	matrix of basis function values
Ψ	matrix of basis function values

References

Adler, D. and D. Murcoch (2009). *rgl: 3D visualization device system (OpenGL).* R package version 0.82. http://rgl.neoscientists.org.

Bellman, R. and R. S. Roth (1971). The use of splines with unknown end points in the identification of systems. *Journal of Mathematical Analysis and Applications 34*, 26–33.

Bookstein, F. L. (1991). *Morphometric Tools for Landmark Data: Geometry and Biology.* Cambridge: Cambridge University Press.

Borrelli, R. L. and C. S. Coleman (2004). *Differential Equations: A Modelling Perspective.* New York: Wiley.

Brumback, B. A. and J. A. Rice (1998). Smoothing spline models for the analysis of nested and crossed samples of curves. *Journal of the American Statistical Association 93*, 961–994.

Brunel, N. (2008). Parameter estimation of ODEs via nonparametric estimators. *Electronic Journal of Statistics 2*, 1242–1267.

Cardot, H., F. Ferraty, A. Mas, and P. Sarda (2003b). Testing hypotheses in the functional linear model. *Scandanavian Journal of Statistics 30*, 241–255.

Cardot, H., F. Ferraty, and P. Sarda (1999). Functional linear model. *Statistics und Probability Letters 45*, 11–22.

Cardot, H., F. Ferraty, and P. Sarda (2003a). Spline estimators for the functional linear model. *Statistica Sinica 13*, 571–591.

Cardot, H., A. Goia, and P. Sarda (2004). Testing for no effect in functional linear models, some computational approaches. *Communications in Statistics— Simulation and Computation 33*, 179–199.

Chambers, J. M. (2008). *Software for Data Analysis.* New York: Springer.

Chambers, J. M. and T. J. Hastie (1991). *Statistical Models in S.* New York: Chapman and Hall.

Chaudhuri, P. and J. S. Marron (1999). SiZer for exploration of structures in curves. *Journal of the American Statistical Association 94*, 807–823.

Chen, J. and H. Wu (2008). Estimation of time-varying parameters in deterministic dynamic models. *Statistica Sinica 18*, 987–1006.

Chiou, J. M. and H. G. Müller (2009). Modeling hazard rates as functional data for the analysis of cohort lifetables and mortality forecasting. *Journal of the American Statistical Association*, in press.

Craven, P. and G. Wahba (1979). Smoothing noisy data with spline functions: Estimating the correct degree of smoothing by the method of generalized cross-validation. *Numerische Mathematik 31*, 377–403.

Cuevas, A., M. Febrero, and R. Fraiman (2002). Linear functional regression: The case of fixed design and functional response. *Canadian Journal of Statistics 30*, 285–300.

de Boor, C. (2001). *A Practical Guide to Splines, Revised Edition.* New York: Springer.

Delsol, L., F. Ferraty, and P. Vieu (2008). Structural test in regression on functional variables. to appear.

Escabias, M., A. Aguilera, and M. J. Valderrama (2004). Principal component estimation of functional logistic regression: Discussion of two different approaches. *Nonparametric Statistics 16*, 365–384.

Eubank, R. L. (1999). *Spline Smoothing and Nonparametric Regression, Second Edition.* New York: Marcel Dekker.

Fan, J. and I. Gijbels (1996). *Local Polynomial Modelling and Its Applications.* London: Chapman and Hall.

Ferraty, F. and P. Vieu (2001). The functional nonparametric model and its applications to spectometric data. *Computational Statistics 17*, 545–564.

Fisher, N. I., T. L. Lewis, and B. J. J. Embleton (1987). *Statistical Analysis of Spherical Data.* Cambridge: Cambridge University Press.

Gasser, T. and A. Kneip (1995). Searching for structure in curve samples. *Journal of the American Statistical Association 90*, 1179–1188.

Gervini, D. and T. Gasser (2004). Self–modeling warping functions. *Journal of the Royal Statistical Society, Series B 66*, 959–971.

Hastie, T. and R. Tibshirani (1993). Varying-coefficient models. *Journal of the Royal Statistical Society, Series B 55*, 757–796.

Hiebeler (2009). Matlab / R reference. http://www.math.umaine.edu/faculty/hiebeler /comp/matlabR.pdf, accessed 2009.02.06.

James, G., J. Wang, and J. Zhu (2009). Functional linear regression that's interpretable. *Annals of Statistics*, in press.

James, G. M. (2002). Generalized linear models with functional predictors. *Journal of the Royal Statistical Society, Series B 64*, 411–432.

James, G. M. and T. Hastie (2001). Functional linear discriminant analysis for irregularly sampled curves. *Journal of the Royal Statistical Society, Series B 63*, 533–550.

James, G. M., T. J. Hastie, and C. A. Sugar (2000). Principal component models for sparse functional data. *Biometrika 87*, 587–602.

James, G. M. and C. A. Sugar (2003). Clustering sparsely sampled functional data. *Journal of the American Statistical Association 98*, 397–408.

Jolliffe, I. T. (2002). *Principal Components Analysis, Second Edition.* New York: Springer.

Kneip, A. and T. Gasser (1992). Statistical tools to analyze data representing a sample of curves. *Annals of Statistics 20*, 1266–1305.

Kneip, A. and J. O. Ramsay (2008). Combining registration and fitting for functional models. *Journal of the American Statistical Association 20*, 1266–1305.

Kuznetsov, Y. A. (2004). *Elements of Applied Bifurcation Theory*. New York: Springer.

Liu, X. and H. G. Müller (2004). Functional convex averaging and synchronization for time-warped random curves. *Journal of the American Statistical Association 99*, 687–699.

Malfait, N. and J. O. Ramsay (2003). The historical functional linear model. *Canadian Journal of Statistics 31*, 115–128.

Müller, H.-G. and U. Stadtmüller (2005). Generalized functional linear models. *Annals of Statistics 33*, 774–805.

Olshen, R. A., E. N. Biden, M. P. Wyatt, and D. H. Sutherland (1989). Gait analysis and the bootstrap. *Annals of Statistics 17*, 1419–1440.

Pascual, M. and S. P. Ellner (2000). Linking ecological patterns to environmental forcing via nonlinear time series models. *Ecology 81*(10), 2767–2780.

Ramsay, J. O., R. D. Bock, and T. Gasser (1995a). Comparison of height acceleration curves in the Fels, Zurich, and Berkeley growth data. *Annals of Human Biology 22*, 413–426.

Ramsay, J. O., G. Hooker, D. Campbell, and J. Cao (2007). Parameter estimation in differential equations: A generalized smoothing approach. *Journal of the Royal Statistical Society, Series B 16*, 741–796.

Ramsay, J. O. and B. W. Silverman (2005). *Functional Data Analysis, Second Edition*. New York: Springer.

Ramsay, J. O., X. Wang, and R. Flanagan (1995b). A functional data analysis of the pinch force of human fingers. *Applied Statistics 44*, 17–30.

Rossi, N., X. Wang, and J. O. Ramsay (2002). Nonparametric item response function estimates with the em algorithm. *Journal of the Behavioral and Educational Sciences 27*, 291–317.

Rupert, D., M. P. Wand, and R. J. Carroll (2003). *Semiparametric Regression*. Cambridge: Cambridge University Press.

Sakoe, H. and S. Chiba (1978). Dynamic programming algorithm optimization for spoken word recognition. *IEEE Transactions, ASSP-26 1*, 43–49.

Sarkar, D. (2008). *lattice: Lattice Graphics*. R package version 0.17-13.

Schumaker, L. (1981). *Spline Functions: Basic Theory*. New York: Wiley.

Silverman, B. W. (1986). *Density Estimation for Statistics and Data Analysis*. London: Chapman and Hall.

Simonoff, J. S. (1996). *Smoothing Methods in Statistics*. New York: Springer.

Tuddenham, R. D. and M. M. Snyder (1954). Physical growth of California boys and girls from birth to eighteen years. *University of California Publications in Child Development 1*, 183–364.

Varah, J. M. (1982). A spline least squares method for numerical parameter estimation in differential equations. *SIAM Journal on Scientific Computing 3*, 28–46.

Yao, F., H.-G. Müller, and J.-L. Wang (2005). Functional data analysis for longitudinal data. *Annals of Statistics 33*, 2873–2903.

Zwiefelhofer, D., J. H. Reynolds, and M. Keim (2008). Population trends and annual density estimates for select wintering seabird species on Kodiak Island, Alaska. Technical report, U.S. Fish and Wildlife Service, Kodiak National Wildlife Refuge. Technical Report, no. 08-00x.

Index

A Beginner's Guide to R

Alain F. Zuur, Elena N. Ieno, Erik H.W.G. Meesters, and Den Burg

The text covers how to download and install R, import and manage data, elementary plotting, an introduction to functions, advanced plotting, and common beginner mistakes.

Content: Introduction.- Getting data into R.- Accessing variables and managing subsets of data.- Simple commands.- An introduction to basic plotting tools.- Loops and functions.- Graphing tools.- An introduction to lattice package.- Common R mistakes.

2009. Approx. 215 p. Softcover (Use R)
ISBN: 978-0-387-93836-3

Functional Data Analysis

J. Ramsay
B. W. Silverman

Content: Introduction .- Tools for Exploring Functional Data .- From Functional Data to Smooth Functions .- Smoothing Functional Data by Least Squares .- Smoothing Functional Data with a Roughness Penalty .- Constrained Functions .- The Registration and Display of Functional Data .- Principal Components Analysis for Functional Data .- Regularized Principal Components Analysis .- Principal Components Analysis of Mixed Data .- Canonical Correlation and Discriminant Analysis .- Functional Linear Models .- Modelling Functional Responses with Multivariate Covariats .- Functional Responses, Functional Covariates and the Concurrent Model .- Functional Linear Models for Scalar Responses .- Functional Linear Models for Functional Responses .- Derivatives and Functional Linear Models .- Differential Equations and Operators .- Principal Differential Analysis .- Green's Functions and Reproducing Kernels .- More General Roughness Penalties .- Some Perspectives on FDA.

2005. 2nd ed. XX, 430 p. 151 illus. Hardcover (Springer Series in Statistics)
ISBN: 978-0-387-40080-8

Nonparametric Functional Data Analysis Theory and Practice

Frédéric Ferraty
Philippe Vieu

Content: Introduction to functional nonparametric statistics.- Some functional datasets and associated statistical problematics.- What is a well adapted space for functional data?.- Local weighting of functional variables.- Functional nonparametric prediction methodologies.- Some selected asymptotics.- Computational issues.- Nonparametric supervised classification for functional data.- Nonparametric unsupervised classification for functional data.- Mixing, nonparametric and functional statistics.- Some selected asymptotics.- Application to continuous time processes prediction.- Small ball probabilities, semi-metric spaces and nonparametric statistics.- Conclusion and perspectives.

2006. XX, 268 p. 29 illus. Hardcover (Springer Series in Statistics)
ISBN: 978-0-387-30369-7